INSECT PATHOLOGY

Textbook and Practical Manual

Nitesh Kumar Maru
Ashwani Kumar
Sunil Zachariah

SCIENTIFIC PUBLISHERS

Published by

SCIENTIFIC PUBLISHERS (INDIA)
5 A, New Pali Road, P.O. Box 91
Jodhpur 342 001 (INDIA)
E-mail: info@scientificpub.com
Website : www.scientificpub.com

ISBN: 978-93-88812-81-8
eISBN: 978-93-88812-82-5

Printed in India

PREFACE

This work is originally destined principally for the use of students enrolled in the course in Insect Pathology being given at the Sam Higginbottom University of Agriculture, Technology and Sciences, Naini, Allahabad. The need of a concise textbook in this subject felt necessary as not only did the students find the literature on the diseases and other pathological conditions of insects, but also no reasonably complete treatment of the subject from the student's standpoint was available. To fill this need is the first objective of this project.

Considering the field as a whole, however, it was realized that, in addition to a textbook for classroom use, the subject was also greatly in need of a reference work. For most of the information the research worker has been forced to consult widely scattered and often almost inaccessible articles or incomplete accounts in larger volumes. In 1946 Edward A. Steinhaus published "Insect Microbiology," purely a reference book, dealing in a general way with the microbiology of insects. Way back in 1949, Edward A. Steinhaus published first edition of "Insect Pathology," which is vast and immensely cover the subject. Since then, so much advancement has been made in the subject that should need to be incorporated. Accordingly, the author kept his attention to the preparation of the present book, which deals specifically with insect pathology, *i.e.,* with the microbial diseases of insects as well as with certain amicrobial diseases, injuries, and abnormalities. The task has not been an easy one. The accomplishments of workers in this field frequently are difficult to evaluate and present in a logical, easily readable form. Also, the field of insect pathology is much larger than most persons realize. Furthermore, the subject matter is peculiarly difficult to present both as a textbook and as a reference book. This nevertheless is what authors have attempted to accomplish.

Although, pertinent references have been cited throughout the book, no attempt has been made to make the volume completely bibliographic. The literature contains over 250 odd references dealing with the diseases of insects. However, care has been taken to list those references which will lead the reader to most of the significant contributions made in each particular phase of the subject.

Chapters 1 and 2, dealing with the introduction and history of insect pathology, have intentionally been held to a minimum, and the emphasis of the book has been placed on the microbial diseases of insects. A substantial portion of Chapters 5 and 7 may be considered as belonging to the fields of insect physiology and insect toxicology, respectively. Nevertheless, the pathologies and abnormalities concerned are too frequently ignored by textbook writers in these fields. Since it is important for purposes of orientation that the student in insect pathology be fortified with information dealing with a microbic pathologies, a brief discussion of the latter is included in the present work, but the treatment given them is not intended to be comprehensive.

A book of this kind, dealing with several large and distinct groups of microorganisms as well as with insects, is likely to be subjected to peccadilloes concerning the synonymies and forms of names. An attempt to avoid them has been made by following the nomenclature used by the leading authorities in the various fields concerned. In spite of constant vigilance, errors probably have crept in; and, with regard to the book as a whole, we can but echo Chaucer's humble petition that "If there is anything that displeases them [the readers], I pray also that they ascribe it to the fault of my lack of skill, and not to my intention, which would gladly have expressed it better if I had had the skill to do so."

To write a scientific book, it is usually necessary first to have an environment in which such work may be prepared and nurtured. Credit for supplying this environment belongs to Professor **Dr. Sunil Zachariah**, of the SHUATS, Allahabad, who, more than anyone else, made possible not only this book but also provided basic idea of Draft. Without the warm encouragement and kind understanding of Professor Zachariah, the writing of this book probably would not have been attempted.

Authors are also greatly indebted to others who have been generous in their help and advice. For critically reading portions of the manuscript, wish to express my deep appreciation to other colleagues at the SHUATS Allahabad. Authors are particularly indebted to **Dr. Ashwani Kumar** for reading the entire manuscript and thankful to **Dr. Anurag R. Tayde** and **Mr. D. Vamsi Krishna** for providing assistance in writing this project warm heartedly.

Thanks to the numerous persons who provided prints of photographic material and permitted their reproduction in this book. Acknowledgments of these courtesies are made at the places where the material has been used. Also wish to thank **Mr. Suryadatt Pandey,** who painstakingly typed the manuscript, and who assisted in much of the work of indexing.

Authors

Allahabad
October 20, 2019

BEFORE WE START...

Second section of this book describes a number of techniques which will be useful to insect pathologists working in the laboratory. Most of the techniques will be familiar to plant or medical pathologists. The reader is strongly advised to seek advice and assistance from their course teacher or specialists trained in this fields. The techniques described herein assume that the pathologist has access to a reasonably well-equipped laboratory, including autoclave or pressure cooker, microscope, glassware, and chemicals.

To study or identify non-occluded viruses you will need access to an electron microscope. For bacteria and occluded viruses, use a compound microscope with an oil-immersion objective lens. Fungi and protozoa can be observed under a compound light microscope. Nematodes can usually be seen with the naked eye.

CONTENT

Theory

1.	Introduction	1-3
2.	History of insect pathology	4-16
3.	Infection of insects by various pathogen	17-32
4.	Concept of Epizootiology	33-45
5.	Defence mechanisms in insects against pathogens	46-55
6.	Mass production techniques of pathogens	56-70
7.	Safety and registration of microbial pesticides in India	71-78
8.	Use of insect pathogens in IPM	79-98

Practicals

1.	Equipment used and safety measures to be observed in insect pathology laboratory	101-107
2.	Identification and symptoms of infection – bacterial insect pathogens	108-111
3.	Identification and symptoms of infection – fungal insect pathogens	112-117
4.	Identification and symptoms of infection – viral insect pathogens	118-122
5.	concept of pathogen load	123-124
6.	Extraction, Isolation and Culturing of fungal pathogens from live or dead insect/organism	125-126
7.	Extraction, Isolation and Culturing of bacterial pathogens from live or dead insect/organism	127-129

8.	Extraction, Isolation and Culturing of viral pathogens from live or dead insect/organism	130-131
9.	To know the different laboratory bioassay techniques for evaluation of different insecticides against insect pests	132-137
10.	Bioassay of Fungal Entomopathogens against Grass hoppers or Locusts	138-143

Chapter 1

INTRODUCTION

The "Insect Pathology" refers to that branch of entomology which deals with the general principles of pathology as they may be applied to insects. In a broad sense, it refers to observations concerning the cause, Symptomatology, and Epizootiology of the diseases of insects, and to a study of the structural, chemical, and functional alterations in the body of the insect resulting from disease or injury. The words "Insect Pathology" may be thought of as referring to insects in a manner similar to that in which "Plant Pathology" refers to plants. (Steinhaus, 1949)

Scope of Insect Pathology

Insects through their diversity in types, numbers, life cycles, and habitats, expose themselves to a large number of pathogens. Whatever goes wrong with an insect may be considered pathology or disease. This aspect will be discussed in details in following chapters of this book.

Insect pathology is a branch of invertebrate pathology. However, its scope and development are much more extensive than other areas of invertebrate pathology because two third of invertebrates are insects. Also, insects are economically and medically more important to human than any other invertebrate; hence a large number of pathologists work in insect pathology than in any other area of invertebrate pathology.

Same as in the case of plant and vertebrate pathologies, the general principal of pathology apply to disease of insect. They involve the etiology (Cause), symptomatology and Epizootiology of disease, and also the structural, chemical and functional alterations that results from disease. In addition to entomology, other sciences, such as virology, bacteriology, mycology, protozoology, nematology, pathology, immunology etc. are closely interrelated in the study of insect disease (Tanada and Kaya, 2012). This means that an insect pathologist should have a very broad background. He or she may be specialized in one or few areas along with a general understanding of the different types of disease in insects.

Interrelation of Insect Pathology with other Branches of Entomology

Perhaps the first apparent application of insect pathology that comes to the mind of the student in entomology is its use in the field of biological control, since insects are subject to diseases just as are other animals, the possibility of controlling serious insect pests by the dissemination of disease organisms among them not only appears plausible, but in several instances has proved to be entirely possible and practicable. Of even greater significance, than the artificial use of these diseases is the very effective control of insects, that takes place in nature through the agency of disease without the help of man.

Insect pathology may be valuable to economic entomology in ways other than that of artificial biological control. The entomologist studying the ecology of insects frequently finds himself depending on the insect pathology for an understanding of disease outbreak before he can correctly interpret his other observations. Those who rear insects in the laboratory or in insectaries are finding that the control of disease in their insects frequently presents problems for which the field of insect pathology may offer solutions.

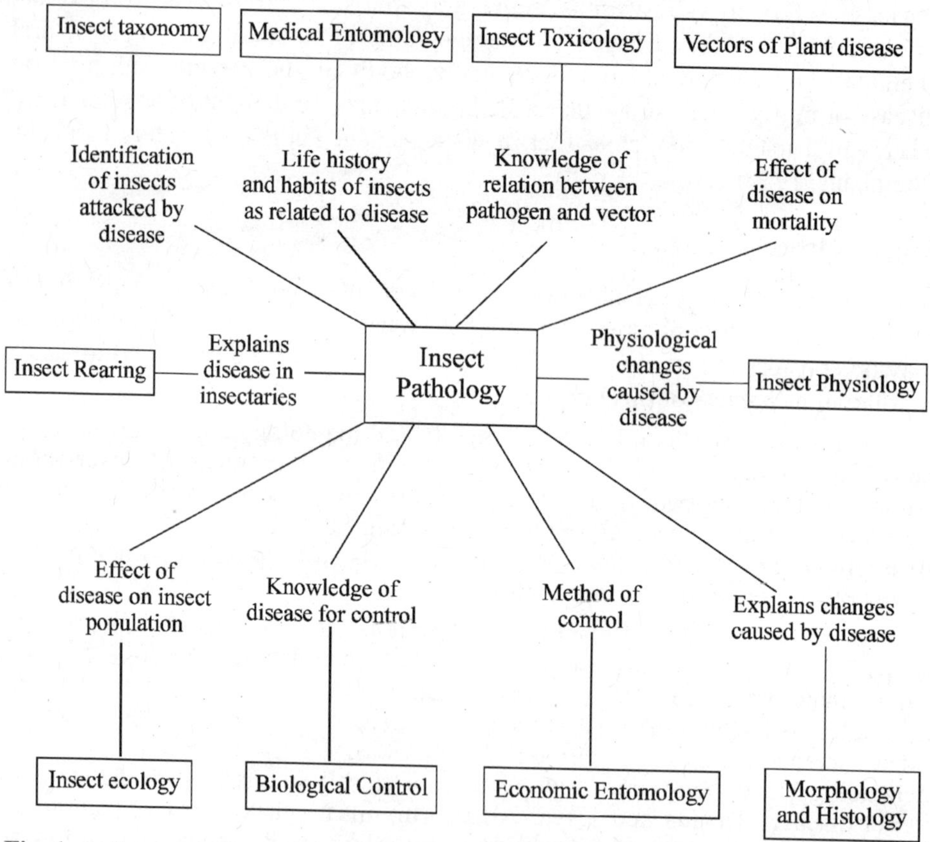

Fig. 1. A diagrammatic representation of the Interrelation of insect pathology with other branches of entomology

The value of insect pathology to medical entomology in particular, and to parasitology generally, is probably fairly obvious, but it is surprising how little correlation is made between the two. One of the basic maxims of insect pathology is the gaining of an understanding of the biological relationships between insects and microorganisms. This is just the type of information and study so greatly needed in medical entomology.

Insect physiology and insect pathology have a great deal in common, each contributing greatly to the development of the other. Diseases of insects always involve a disruption of the metabolism and other physiological processes of an insect, and hence the knowledge gained in the study of these diseases may be in the form of direct contributions to insect physiology.

A prerequisite to a thorough understanding of the principles of insect pathology is a detailed knowledge of the insect anatomy and cellular structure of insects. Similarly, insect pathologists have found it necessary in many cases to make careful comparative studies of the blood structure and modification, therein diseased and normal insects; these contributions should be added to similar ones being made by the insect physiologist.

Since the full meaning of the term "insect pathology" includes certain aspects of insect toxicology, one need only point out that, when insects are sickened, killed, or otherwise affected by toxic chemical agents, a pathological state results.

Now, just as insect pathology can be useful to other branches of entomology, the converse is true. Insect pathology is absolutely dependent upon every other branch of entomology, excluding none (Fig. 1). Not the least of these dependencies is that in which the insect pathologist relies upon the field entomologist to bring him reports of disease outbreaks, as well as to send in diseased specimens collected in nature. In the laboratory, the insect pathologist is dependent upon the techniques and procedures of all the various branches of entomology.

References

Steinhaus, E.A. (1949). Principles of insect pathology.

Steinhaus, E.A. ed. (2012). *Insect Pathology V1: An Advanced Treatise* (Vol. 1). Elsevier

Tanada, Y. and Kaya, H.K. (2012). *Insect Pathology*. Academic press.

Question Time

1. Define insect pathology. Discuss in brief its scope and importance.
2. How is insect pathology related to other branches of entomology? Discuss in detail with appropriate diagram.
3. Discuss the importance of insect pathology in economic entomology.

One Word Answer

1. Father of modern insect pathology – *Edward Steinhaus.*
2. Branch of entomology which deals with the general principles of pathology as they may be applied to insects – *Insect Pathology.*
3. In addition to entomology, other sciences *viz.* virology, bacteriology, mycology, nematology, etc. are closely interrelated in the study of – *Insect Disease.*

■■■■

Chapter 2

HISTORY OF INSECT PATHOLOGY

Insects get sick too. Anyone working with insects in the field or in a laboratory will come across insects that are infected by pathogens. The field of insect pathology tries to identify the pathogens and quantify their effect on individual insects, and on insect populations. Insect pathology is not a new scientific area. Below are mentioned some of the major historical events that shaped the field of insect pathology.

Timeline	Events Influencing Insect Pathology
6000 BC	Humans harvest honey from honey bees

The earliest reports of insect disease in WESTERN Literature were all honey bee and silkworm diseases, reported as far back as 700 BC, and received fair mention in:

"Myth of Aristaeus" (700 BC)

Aristaeus, apparently the discoverer of beekeeping lost his hives to disease. He compelled Proteus, the wise God of the sea to tell him how to avoid such a loss in the future.

Aristaeus

Aristotle

Aristotle (322 BC) in:

"Historia Animalium" (322 BC)

Mentionsed visible parasites and other ailments of insects, such as ants, flies and grasshoppers

Timeline	Events Influencing Insect Pathology
37–39 BC	*Virgil* records honey bee diseases
77 AD	*Pliny* records honey bee diseases

555 AD	Silkworm was introduced to Europe and Middle East
1527	*Marco Girolamo Vida* publish the poem "*De Bombyce*"
1602	*Aldrovandi* described worms emerging from grasshoppers

Illustration made by M.S. Merian

Maria Sibylla Merian

In **1679** Maria Sibylla Merian (1647–1717) describes **"Melted" caterpillars**. Maria was a naturalist and scientific illustrator. She is also sometimes considered the "Mother of Entomology". Her most pivotal work, **"Metamorphosis insectorum surinamensium"** Contains illustrations and natural history observations of the insects that she observed during her trip to **`Suriname'** (Country in South America).

Timeline	Events Influencing Insect Pathology
1734–1742	*Re´amur* finds Cordyceps in silkworms, nematodes in bumble bees
1747	*Gould* describe nematodes in ants
1758	*Linnaeus* describe nematodes in invertebrates in *Systema Naturae*
1771	*Schirach* describe foulbrood disease of honey bees
1808	*Nysten* publish the first book on silkworm diseases
1808–1813	*Bassi* studied fungus disease of silkworms

William Kirby

William Spence

1822-William Kirby (left): He is often considered as the English "Father of Entomology" and is well known for his description and collection of beetles. An enlightened Mr. Kirby also divided diseases between those that resulted from injury or *'internal dearrangement'* and those that were caused by parasitic agents.

1822-William Spence (right): While not as prolific as Kirby, Spence was a well-respected entomologist and was one of the founding members of the Society of Entomologists of London.

In 1822, Kirby and Spence published one of the first popular English entomology books

"An Introduction to Entomology or Element of the Natural History of Insects", in which they devote a chapter to the Diseases of Insects.

In **1835** in Italy Agostino Bassi publishes his discovery of *Beauveria bassiana* in **"Del mal del segno, Calcinaccio o Moscardino"**.

Agostino Bassi is often forgotten, but he played an important role in our understanding of disease. He was the first to demonstrate the **germ theory of disease**. He was the first to understand the etiology of a microbial disease while he worked with muscardine disease in silk worms which was causing severe economic hardship in that silkworm industry in Italy and France at that time. He is often considered the **"Father of Insect Pathology".**

Agostino Bassi

Timeline	Events Influencing Insect Pathology
1855	*Cohn* describe *Empusa* fungus in house fly
1857	*Naegeli* describe *Nosema bombycis*, from silkworms
1870	*Pasteur* published *Etudes sur la Maladie des Vers a Soie*
1873	*LeConte* proposed microbial control to American Association for the Advancement of Science meeting
1883	*Sorokin* named green muscardine *Metarrhizium anisopliae*
1885	*Cheshire* and *Cheyne* showed that foulbrood disease in honey bee was caused by *Bacillus alvei*

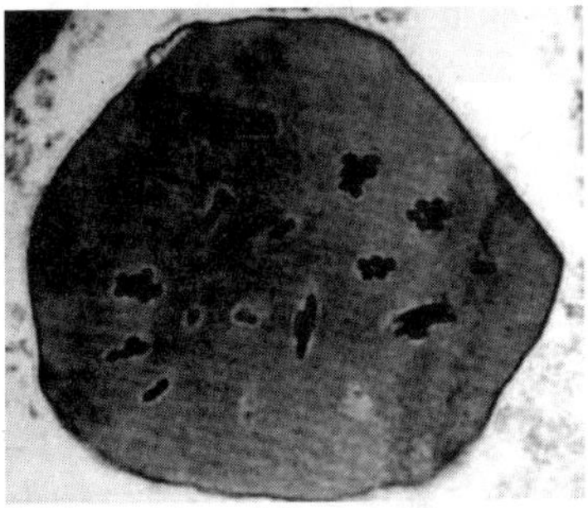

Microscopic view of NPVs

The Italian scientists **Maestri and Coelia in 1856** first observed Nucleopolyhedrovirues (NPVs). They were the first to establish the correlation between the presence of refractive crystals (viral occlusion bodies) and the symptoms of "jaundiced" silkworms.

Louis Pasteur, was a chemist by training, but during an outbreak, of a disease in 1865, that threatened to topple the French silk industry, he was recruited to study the cause of the disease and potentially eradicate it.

Louis Pasteur

The disease was Pebrine. Pasteur discovered that Pebrine was caused by the microsporidian *Nosema bombycis*. An organism that was first described by **Nageli in 1857**.

Pasteur demonstrated that this disease could be passed by contact, contaminated food, and vertically from mother to offspring through the egg. This was the first demonstration of **vertical transmission** of a pathogen.

John Lawrence LeConte

In **1873 John Lawrence LeConte,** proposed to use insects and their diseases to study mechanisms of disease transmission. He also proposed that the muscardine fungus could be useful in controlling pest caterpillar species.

John L. LeConte was one of the most important early American Entomologists. He is responsible for describing a large majority of North American insects' species, most notably in order Coleoptera.

Élie Metchnikof

In **1878** Russian zoologist **Élie Metchnikof** found a green muscardine fungus, *Entomophthora anisopliae.* This fungus was later named *Metarhizium anisopliae* by **Sorokin (1883)**. Metchnikoff found that conidia contaminated soil could infect larvae.

He was also one of the pioneers to suggest that a pathogen could be used as method of pest control. Élie Metchnikof who won the **Nobel Prize Physiology**

or Medicine in **1908** with **Paul Ehrlic** for studies of phagocytosis, inflammation and immunity, ideas that begun with his work on insect disease.

Timeline	Events Influencing Insect Pathology
1887	*Institut Pasteur* (Pasteur Institute) was founded
1888	*Krassilstchik* attempted biocontrol using *Metarhizium anisopliae*
1888	*Lugger* and *Snow* attempted control of chinch bug using *B. bassiana*
1894	*Bolle* observed NPV polyhedra soluble in gut juices

Stephen Forbes

In **1898, Stephen Forbes** (First Head of the Department of Entomology, of University of Illinois at Urbana-Champaign, USA and also Head of the Illinois Natural History Survey), described a polyhedrosis virus from the true army worm *Pseudaletia unipuncta* Haworth or *Mythimna unipuncta* H. (1809)

Microscopic view of bacterium *Bacillus thuringiansis* (Bt.)

In **1901** Japanese biologist **Ishiwatari** was investigating the cause of the **sotto disease** (sudden-collapse disease) that was killing large populations of silkworms. When he first isolated the bacterium *Bacillus thuringiensis* (Bt), found it to be the cause of the disease. In **1911,** the German researcher **Berliner** isolated a bacterium that was killing the Mediterranean flour moth and rediscovered Bt. He named it *Bacillus thuringiensis*, after the German town Thuringia, where the moth was found. (Ishiwatari had named the bacterium *Bacillus sotto* in **1901** but the name was later ruled invalid).

Timeline	Events Influencing Insect Pathology
1906	*G.F. White* found two foulbrood diseases in honey bees
1909	*Zander* described *Nosema* apis from honey bees
1909	*Berliner* discovered *Bacillus thuringiensis* in flour moth
1911	*d'Herelle* observed bacterial epizootics in grasshoppers
1912	*White* found European foulbrood caused by *Bacillus pluton*
1912	*Fantham* and *Porter* described structure of *Nosema apis*
1913, 1916	*Maassen* described chalkbrood in honey bees
1913	*Glaser* and *Chapman* demonstrated that `wilt disease' in gypsy moth is a viral disease
1915	*Berliner,* the existence of crystals within *Bacillus thuringiensis*
1917	*F.G. White* described the first virus in honey bees
1921	*Keilin* described first fungus in aquatic insect *(Coelomomyces in mosquitoes)*
1923	*Steiner described first species of entomopathogenic nematode [Neoaplectana (Steinernema) kraussei]*
1929	*Bacillus thuringiensis* and *Metarhizium* discussed at International Corn Borer Conference in Paris
1934	*Filipjev* suggested use of nematodes for biocontrol
1935	First release of *Steinernema glaseri* against *Japanese beetles*
1938	First commercial *B. thuringiensis* product *(Sporeine)* marketed in France
1953	*Canning* described *Nosema locustae*
1956	*Steinhaus* published "*Living insecticides*" in Scientific American

Edward Steinhaus (1914–1969)

Edward Steinhaus (1914–1969) is often considered the "**Father of Modern Insect Pathology"**. His first contribution to insect pathology was his PhD dissertation which was published as the "**Catalogue of Bacteria Associated Extracellularly with Insects and Ticks" in 1942**.

He taught the first course on Insect Pathology at the University of California at Berkeley in 1947. He also wrote key textbooks in Insect Pathology including: **Insect Microbiology** (1946), and **Principles of Insect Pathology** (1949).

Steinhaus was a founder of **"The Society for Invertebrate Pathology".**

He demonstrated the effectiveness of *Bacillus thuringiensis* for controlling pest insects.

He was also involved in the first practical application of insect viruses for biological control.

Timeline	Events Influencing Insect Pathology
1956	First International Conference on Insect Pathology and Biological Control at Montreal, Canada
1959	Journal of Insect Pathology first published by Academic Press, Cambridge
1961	First *Bacillus thuringiensis* products registered in USA (Thuricide, Biocontrol)
1962	*Kurstak* isolated *B. thuringiensis* subspecies *kurstaki*
1963	*Vago* described entomopoxviruses
1963	*Tanada* found and described granulovirus in codling moth

Timeline	Events Influencing Insect Pathology
1963	*Steinhaus* published Insect Pathology: An Advanced Treatise
1965	*Kellen et al.* reported pathogenic *Bacillus sphaericus* in mosquitoes
1965	Journal of Insect Pathology renamed Journal of Invertebrate Pathology
1966	*Huger* discovered virus in rhinoceros beetle
1967	Society for Invertebrate Pathology was established
1973	*Helicoverpa zea* NPV registered, first virus product (Viron/H, Elcar)
1976	*Goldberg* and *Margalit* discovered *Bacillus thuringiensis israelensis* pathogenic to mosquitoes
1978	Gypchek NPV registered against gypsy moth
1980	*Nosema locustae* approved for control of grasshoppers
1981	*Hirsutella thompsonii* registered for use against citrus mite in USA
1981	*Schnepf* and *Whitely* reported 1st Bt. toxin gene sequence
1981	*Bacillus thuringiensis israelensis* registered against mosquitoes
1983	*Mullis* developed polymerase chain reaction (PCR)
1987	First transgenic plants with *Bacillus thuringiensis* genes reported
1995	Transgenic plants with *Bacillus thuringiensis* genes marketed

Gypsy moth larvae　　Hardwood damage in forest

In **1978 Gypchek NPV** registered against gypsy moth (*Lymantria dispar dispar* L.). One of the most destructive pests of hardwood trees in the eastern United States.

In **1981** *Bacillus thuringiensis israelensis* registered for use against mosquitoes.

Colorado potato beetle

In **1988** ***Bacillus thuringiensis tenebrionis*** registered for use against Colorado potato beetle in USA.

Timeline	Events Influencing Insect Pathology
2007	*Bacillus thuringiensis* genome published by *Challacombe et al.*
2011	Genomes of *Metarhizium anisopliae* and *Metarhizium acridum* published by *Gao et al.*

Conclusions

The long history of insect pathology, from the observations of Aristotle more than 2000 years ago to the many useful products derived from insect pathogens today, gives us an appreciation of the potential utility of these organisms. However, it is important to remember that these developments do not happen on their own, but are the result of the hard work and imagination of scientists around the world studying diseases of insects, only a small percentage of which are mentioned here. With the rapid development of molecular methods, many insect pathogens have been fully sequenced, beginning with the black beetle virus in 1985 (Dasgupta *et al.,* 1984; Dasmahapatra *et al.,* 1985) and the first NPV (*Autographa californica*) in 1994 (Ayres *et al.,* 1994). Nematode and insect symbionts, Photorhabdus and Wolbachia, were fully sequenced in 2003 and 2004, respectively (Duchaud *et al.*, 2003; Wu *et al.,* 2004), and the *B. thuringiensis* genome was published in 2007 (Challacombe *et al.,* 2007). The genomes of two bee pathogens, the fungus *Ascosphaera apis* and the bacterium *Paenibacillus larvae*, were published in 2006 (Qin *et al.,* 2006) and the genome of the microsporidian bee pathogen, *Nosema ceranae*, was published in 2009 (Cornman *et al.,* 2009). The genomes of *Metarhizium anisopliae* and *Metarhizium acridum* were published in 2011 (Gao *et al.,* 2011). With the costs of sequencing becoming more affordable every year, the day will come when many insect pathogens will be fully sequenced. This will allow for a better understanding of the evolution of pathogenicity, the presence of virulence genes, metabolite biosynthesis, and genetic relatedness with other organisms and, perhaps, the development of more effective delivery methods in the field.

References

Alleyne, M. and Thairu, M. (2011). Webpage "History of Insect Pathology" DOI: https://www.sutori.com/story/history-of-insect-pathology

Ayres, M.D., Howard, S.C., Kuzio, J., Lopez-Ferber, M. and Possee, R.D. (1994). The complete DNA sequence of *Autographa californica* nuclear polyhedrosis virus. *Virology*, 202: 586–605.

Challacombe, J.F., Altherr, M.R., Xie, G., Bhotika, S.S., Brown, N., Bruce, D., Campbell, C.S., Campbell, M.L., Chen, J., Chertkov, O. and Cleland, C. (2007). The complete genome sequence of *Bacillus thuringiensis* Al Hakam. *Journal of bacteriology*, 189(9): 3680–3681.

Cornman, R.S., Chen, Y.P., Schatz, M.C., Street, C., Zhao, Y., Desany, B., Egholm, M., Hutchison, S., Pettis, J.S., Lipkin, W.I. and Evans, J.D. (2009). Genomic analyses of the microsporidian *Nosema ceranae*, an emergent pathogen of honey bees. *PLoS pathogens*, 5(6): p.e1000466.

Dasgupta, R., Ghosh, A., Dasmahapatra, B., Guarino, L.A. and Kaesberg, P. (1984). Primary and Secondary structure of black beetle virus RNA2, the genomic messenger for BBV coat protein precursor. *Nucl. Acids Res.*, 12: 7215–7223.

Dasmahapatra, B., Dasgupta, R., Ghosh, A. and Kaesberg, P. (1985). Structure of the black beetle virus genome and its functional implications. *J. Molec. Biol.*, 182: 183–189.

Davidson, E.W. (2012). History of insect pathology. In *Insect Pathology* (Second Edition) (pp. 13–28).

Duchaud, E., Rusniok, C., Frangeul, L., Buchrieser, C., Givaudan, A., Taourit, S., Bocs, S., Boursaux-Eude, C., Chandler, M., Charles, J.F., Dassa, E., Derose, R., Derzelle, S., Freyssinet, G., Gaudriault, S., Me´digue, C., Lanois, A., Powell, K., Siguier, P., Vincent, R., Wingate, V., Zouine, M., Glaser, P., Boemare, N., Danchin, A. and

Kunst, F. (2003). The genome sequence of the entomopathogenic bacterium *Photorhabdus luminescens. Nat. Biotechnol.*, 21: 1307–1313.

Gao, Q., Jin, K., Ying, S.H., Zhang, Y., Xiao, G., Shang, Y., Duan, Z., Hu, X., Xie, X.Q., Zhou, G. and Peng, G. (2011). Genome sequencing and comparative transcriptomics of the model entomopathogenic fungi *Metarhizium anisopliae* and *M. acridum. PLoS genetics*, 7(1): e1001264.

Qin, X., Evans, J.D., Aronstein, K.A., Murray, K.D. and Weinstock, G.M. (2006). Genome sequences of the honey bee pathogens *Paenibacillus larvae* and *Ascosphaera apis. Insect molecular biology,* 15(5), pp.715–718.

Steinhaus, E. ed. (2012). Insect Pathology V1: An Advanced Treatise (Vol. 1). Elsevier.

Tanada, Y. and Kaya, H.K., 2012. Insect Pathology. Academic Press.

Wu, M., Sun, L.V., Vamathevan, J., Riegler, M., Deboy, R., Brownlie, J.C., McGraw, E.A., Martin, W., Esser, C., Ahmadinejad, N., Wiegand, C., Madupu, R., Beanan, M. J., Brinkac, L.M., Daugherty, S.C., Durkin, A.S., Kolonay, J.F., Nelson, W.C., Mohamoud, Y., Lee, P., Berry, K., Young, M.B., Utterback, T., Weidman, J., Nierman, W.C., Paulsen, I.T., Nelson, K.E., Tettelin, H., O'Neill, S.L. and Eisen, J.A. (2004). Phylogenomics of the reproductive parasite *Wolbachia pipientis* Mel: a streamlined genome overrun by mobile genetic elements. *PLoS Biol.*, 2: 327–341.

Question Time

1. Discuss in brief the history of insect pathology?
2. Illustrate in brief the contribution of Edward Steinhaus in evolution of modern insect pathology?

One Word Answer

1. Maria Sibylla Merian (1647–1717) describes "melted" caterpillars. She was also sometimes considered as the - "Mother of Entomology".
2. William Kirby is often considered the - "Father of Entomology".
3. Agostino Bassi was the first to demonstrate the "Germ theory of disease", He is often considered as the - "Father of Insect Pathology".
4. Edward Steinhaus wrote key textbooks in Insect Pathology including Insect Microbiology (1946), and Principles of Insect Pathology (1949), often considered as – "Father of modern insect pathology".

Fill in the Blanks

1. *Bacillus thuringiensis* genome published by Challacombe *et al.* In the year – 2007
2. The genome of *Metarhizium anisopliae* published by Gao *et al.* In year - 2011
3. Society for Invertebrate Pathology is established in year – 1967
4. Journal of Insect Pathology first published in – 1959

5. Pasteur Institute was founded in year – 1887
6. First demonstration of vertical transmission of a pathogen- Louis Pasteur
7. Chalk brood disease in honey bees caused by fungus – *Ascosphaera apis*
8. Who suggested use of *nematodes* for bio control of insects – *Filipjev.*

■■■■

Chapter 3

INFECTION OF INSECT BY VARIOUS PATHOGENS

It is now an established fact that Insects are subject to infectious diseases which can seriously affect host fitness. This fundamental attribute is effectively been utilized as applied aspects of insect pathology, since last 2 decades, but surprisingly little is known about the molecular mechanisms of infection (Drion and Pendland, 1998). Historically, economically motivated studies of sicknesses in domesticated insects, such as silkworm or honeybees, have dominated the field. For example, the French government commissioned Louis Pasteur to find a cure for a silkworm epizootic that was threatening the silk industry in 1865. Pasteur managed to identify the parasite that was causing the epidemic (*Nosema bombycis*, a microsporidia that is found in the haemolymph and is transmitted vertically) and his recommendations saved the silk industry. Another factor that has motivated studies of entomopathogenic microorganisms is the need to develop strategies that are based on microbial pathogens to eradicate pest insects. (Federici *et al.*, 2003).

3.1. Infection in Insect Host

The infection process extends from the germination or multiplication of an infective propagule within or on the surface of a potential insect host to the time till pathogen establishes some form of parasitic relationship with its host. The nature of the host-parasite interaction established is determined by the genotypes of the host and the pathogen, and is influenced by the environment. It is convenient to divide the infection process into the three phases of pre-entry (or pre-penetration), entry and colonisation. If the infection process is inhibited at any of these stages of development, a parasitic relationship with the host will not be established and disease symptoms will not develop.

3.2. Site of Pathogen Entry

Insect pathogens may gain entry into insects by one or more major routes. Many opportunistic pathogens gain entry when the host's barriers are compromised. The virus agent responsible for the hairless black syndrome in honeybees, gains entry through breaks in the cuticle caused by removal of spines. Gram negative bacteria may enter through lesions caused by invading pathogens, parasitoids, or predators, resulting in a lethal sepsis. So many pathogen groups are actively transmitted into host insects. The bacteria belonging to the genera *Photorhabdus* and Xenorhabdus are transmitted into insects by their respective nematode hosts. These nematode vectors have been reported to gain access through the mouth, anus, spiracles, and/or the inter segmental domains of the exoskeleton. Certain viruses such as the Poly-DNA-Virus and Asco-viruses are actively and passively transmitted, respectively, by hymenopteran parasitoids to lepidopteran larvae. Several pathogens, certain non-occluded viruses (Rhabdoviruses), protozoans, and gram negative bacteria (*Wolbachia*) are gamete-transmitted and infect female progeny.

The respiratory system, a primary route for the ingress of vertebrate pathogens, is not a substantial portal for the ingress of insect pathogens. Several reports, however, have suggested that certain fungi and nematodes can perforate through the spiracles.

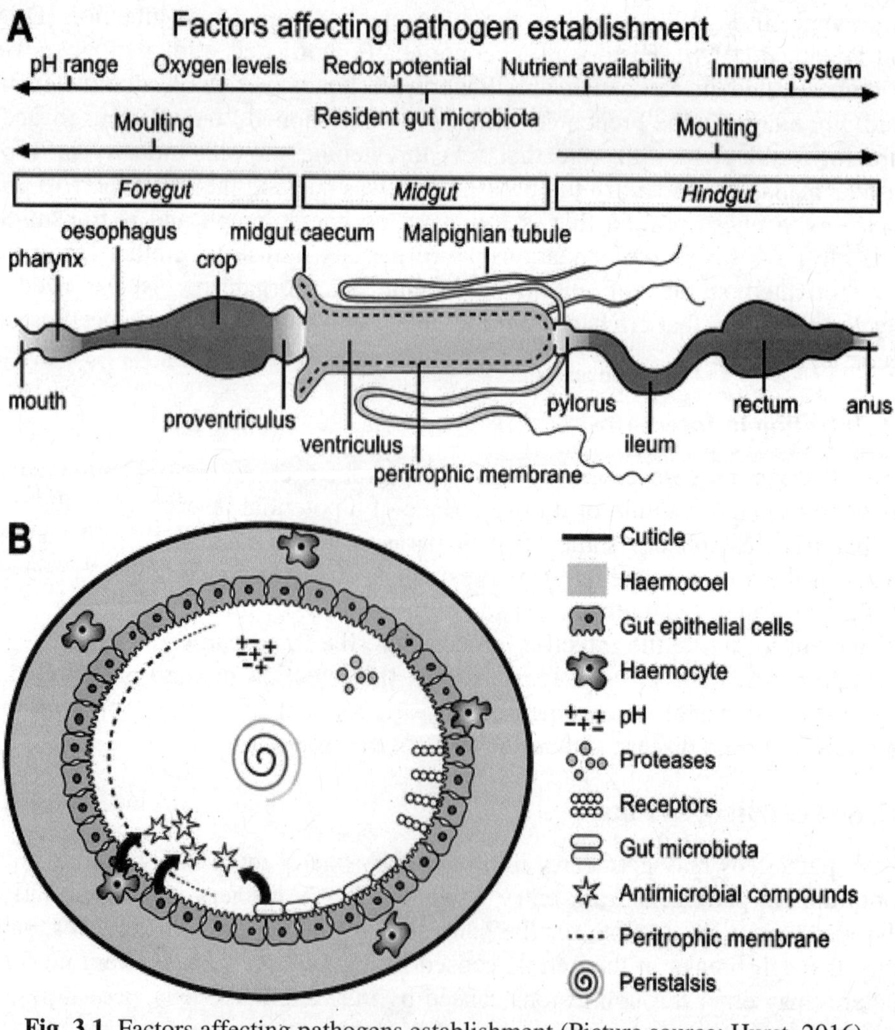

Fig. 3.1. Factors affecting pathogens establishment (Picture source: Hurst, 2016).

The huge majority of the known insect pathogens own the mechanisms to attach to and to enter either the insect cuticle or alimentary tract. Typically, the insect mycopathogens (Fungal pathogens) enter through the cuticle, whereas infectious viral, bacterial, and protozoan pathogens enter through the midgut. The mechanisms by which these disease agents adhere to and penetrate these barriers are considered to be the primary determinants that dictate their specificity and infectivity to insect hosts (Fig. 3.1).

3.3. Role of Insect Integuments to Prevent the Infection

The physical and chemical properties of the insect cuticle produce a highly effective barrier against microbes (Fig. 3.2). The lipids of the epicuticle produces a hydrophobic environment that prevents colonization by many microbes. In several cases the quiescent insect stages (pupae, diapausing insects) produces increased levels of cuticle lipids (oxygenated compounds) capable of suppressing microbial growth (Buckner *et al.,* 1996). Secondly, the predominant cuticle hydrocarbons are the straight chain alkanes, the unsaturated alkenes, and branched alkanes. None of these hydrocarbon groups are readily metabolized by microorganisms. Additionally, certain cuticle lipids such as short chain fatty acids, aldehydes, wax esters, ketones, and alcohols may possess anti-microbial activity (Table 3.1).

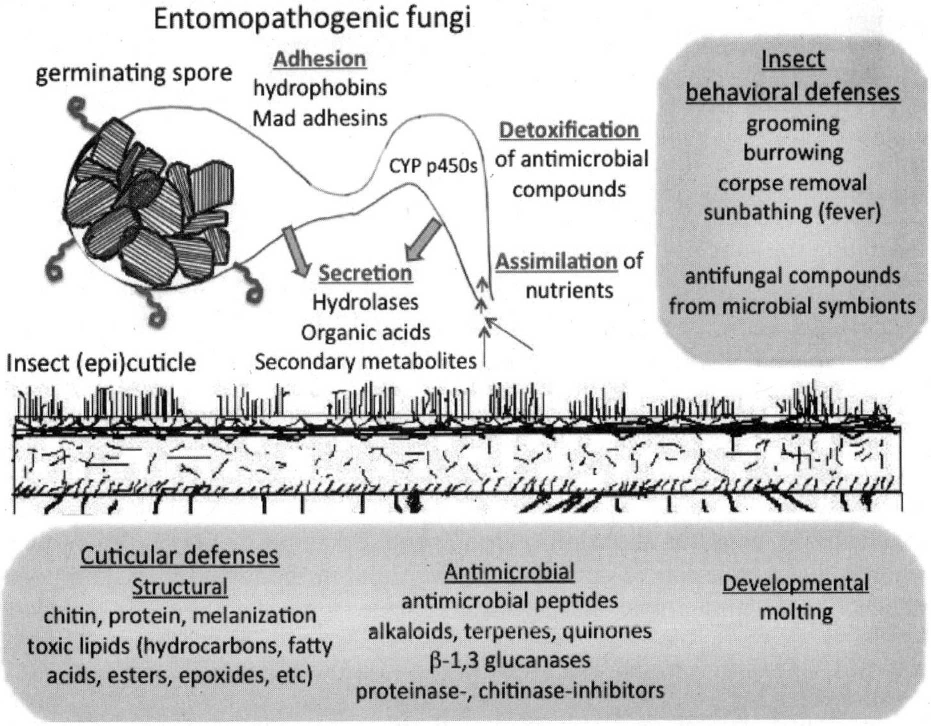

Fig 3.2. Overview of surface interactions between entomopathogenic fungi and the insect cuticle and host behaviours (Source: Ortiz-Urquiza and Keyhani, 2013).

Table 3.1. Antimicrobial substances associated with the insect cuticle.

Compound	Insect
Phenolic groups, quinones	Various insects
Aldehyde (E-2 decanol)	Stinkbugs
Venom alkaloids	Ants
Short-chain fatty acids	Various insects
Cationic peptides	Lepidoptera
Molting fluid	Various insects

The underlying cuticulin layer is heavily sclerotized and contains various phenolic groups that are considered to possess anti-microbial activities. The exocuticle, like the cuticulin layer, is also heavily cross-linked by the sclerotization process. Both of these layers are extremely stable and are resistant to enzymatic and solvent treatment.

Pathogens that penetrate the cuticle often gain access at sites lacking a well-developed exocuticle. Internal and external factors may modulate the biosynthetic activities of the epithelium that in turn alters the physical and chemical properties of the cuticle (Ashida and Brey, 1995).

One of the most prominent changes in cuticle structure occurs at molting, a process which allows for both growth and metamorphosis (Willis, 1996). The effect of insect molting on pathogens has not been extensively analysed. Certain studies suggest that the availability of the new soft cuticle provides an opportunity for access by pathogens incapable of penetrating the sclerotized cuticle (Andersen *et al.,* 1996).

Alternatively, the molting process, which can be stimulated by cuticle damage, may serve as a mechanism to cleanse the insect of potential pathogens inhabiting the cuticle surface.

Wounded Cuticle may be a potential route of entry for pathogens. However, histological studies revealed that cuticle wounding may result in the recruitment of certain haemocytes that clump together in a coagulation response. Recently, it has been demonstrated that the cuticle epithelial cells may contribute to the innate response against microbial invasion.

Case study - Brey *et al.* (1993) reported that mild abrasion of epicuticle followed by topical application of bacteria stimulated epithelial cells to synthesize mRNA transcripts coding for the antibacterial peptide *cecropin*. These observations suggest that a transcuticle signaling event results in the localized synthesis and deposition of antimicrobial substances.

3.4. Role of Insect Alimentary Tract to Prevent the Infection

The majority of infectious bacterial, viral, and protozoan pathogens gain entry into the host through the digestive tract. These disease agents, which contaminate the food substrate, are ingested by insects. The microbes are passively moved with food material through the insect stomodeum (foregut) into the mesenteron (midgut). The ability of an ingested pathogen to interact with the surface of the alimentary tract may be influenced by the flow rate of ingested material. Increased feeding may reduce the chances of microbes to replicate and/or to encounter appropriate host receptors. In certain insects, portions of the anterior digestive tract may act as physical barriers to potential pathogens (Fig. 3.3). For example, the buccal cavity of various ants is modified as a filtration device that prevents or reduces the ingestion of micron-sized particles (Glancy *et al.*, 1981). In general, only a selected group of insect pathogens is capable of colonizing the foregut region.

Case study 1 - The mycoparasitic *Trichomycetes* sp. produce a specialized holdfast structure that attaches the fungus to the cuticle of the foregut of insects.

Case study 2 - The bacterium *Serratia entomophila* is believed to bind the cuticle surface via specialized fimbriae, resulting in colonization and development of amber disease in scarab larvae.

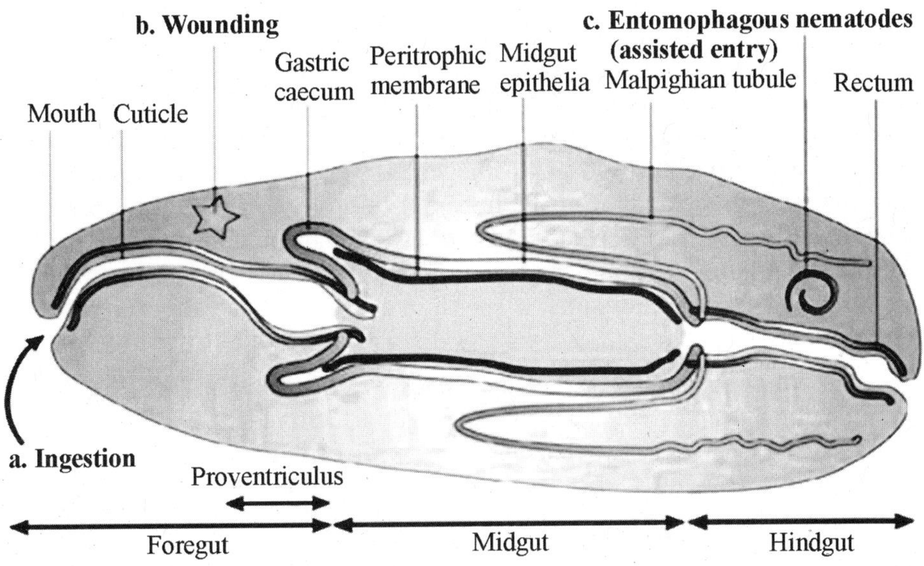

Fig. 3.3. Summary of insect pathogen entry strategies and model of insect gut (*Ref. Vallet-Gely, V., Lemaitre, B. & Boccard, F. Bacterial Strategies of Overcome Insect Defences. Nature Reviews: Microbiology. 6, 302–313 (2008)).*

The mesenteron (midgut) region, unlike the stomodeum (foregut) and proctodeum (hindgut), is not lined with cuticle, is of endodermal origin, and represents one of the major sites for entry of disease agents. The insect midgut has multiple functions, including the digestion and assimilation of ingested food, absorption and secretion of water, and regulation of ion balance. Morphologically, the midgut is a tube that may contain, at its terminal ends, blind pouches or gastric caecae which may harbour various commensalistic microbes. Histological examination of the mid gut reveals that it is comprised of an epithelial layer bordered by the peritrophic membrane on its luminal face and by a basement membrane on the opposite side. A connective tissue sheath traversed by trachea, longitudinal and circular muscles, and nerves encases the basement membrane, segregating it from the insect hemocoel. The visceral muscles, modulated by various aminergic and peptidergic compounds, are responsible for food movement. The epithelial layer underlying the basement membrane is composed of three major cell types: the columnar cells, regenerative cells, and goblet cells. The columnar cells, the major cell type in the midgut, synthesize the digestive enzymes and absorb nutrients. Regenerative cells act as stem cells, providing a source of undifferentiated cells that become incorporated into the midgut epithelium. These cells play an important role in replacing damaged midgut cells. The goblet cells function to transport potassium ions from the haemolymph to the lumen. In several insects, various types of endocrine cells have been detected in histological sections of midgut tissue (Ras and Verbeke, 1994). The peritrophic membrane secreted by midgut cells produces a barrier between the epithelial layer and food bolus (midgut lumen).

3.4.1. Midgut lumen

Ingested microorganisms are faced with multiple physical and chemical barriers in the mesenteron (midgut). The midgut lumen, the site of food digestion, possesses physiochemical properties that may inhibit or kill ingested microbes. In order for a microbe to be a gut pathogen, it must survive the gut pH, redox potential, and ionic strength of the lumen. These physiochemical properties are determined by the insect species, by the developmental stage, and by the food substrate. Furthermore, these properties can vary among the different regions of the alimentary tract, providing a mosaic of favourable and/or unfavourable microhabitats.

The type and function of the digestive enzymes reflect the insect's nutritional habits and are influenced by the gut pH and redox potential (Terra and Ferriera, 2012). The digestive enzymes are produced in and secreted from midgut columnar cells.

The physiochemical and biological properties of the digestive tract often play a key role in the infection process of insect pathogens. Many of the occluded viruses such as the baculovirus, entomopoxvirus, and cytoplasmic polyhedrosis virus require alkaline gut conditions.

The chemical bonds responsible for the crystalline inclusions are disrupted by alkali, resulting in the release of infectious virus.

Case study - Certain bacterial pathogens, such as the causal agent of milky disease, *Bacillus popilliae*, are able to initiate disease after the ingested spores are activated by exposure to digestive proteases. Resulting vegetative cells are then able to initiate the infection process in host scarab larvae.

Digestive enzymes therefore play a dual role in regulating the pathogen/ insect interaction. In certain cases, the hydrolytic enzymes may attack and deactivate pathogens. Alternatively, the lack of appropriate enzymes may preclude the processing of the ingested pathogen.

In addition to the host-derived components, many other factors in the food bolus may affect the infection process. For example, the insect digestive tract, like the vertebrate intestinal tract, may harbour commensalistic and symbiotic microorganisms. The gut microbiota may be antagonistic/inhibitory to insect pathogens.

Case study 1 - The gut microbiota of certain termite species produce volatile fatty acids which may be responsible for the Anti-mycotic activity observed in the subterranean termite *Reticulitermis flavipes* (Boucias *et al.,* 1996).

Case study 2 - Many species of saprophytes associated with the honeybee gut produce Anti-mycotic components active against *Ascosphaera apis*, the causal agent of chalkbrood disease (Bailey, 1981).

Case study 3 - The gram negative bacterium *Enterobacter agglomerans*, common to the desert locust *Schistocerca gregaria*, was found to produce various phenols that possess mycostatic activity (Dillon and Charnley, 1991).

The gut microbiota, in addition to inhibiting ingested pathogens, may serve as microbial antagonists outside the insect host (Fig. 3.4).

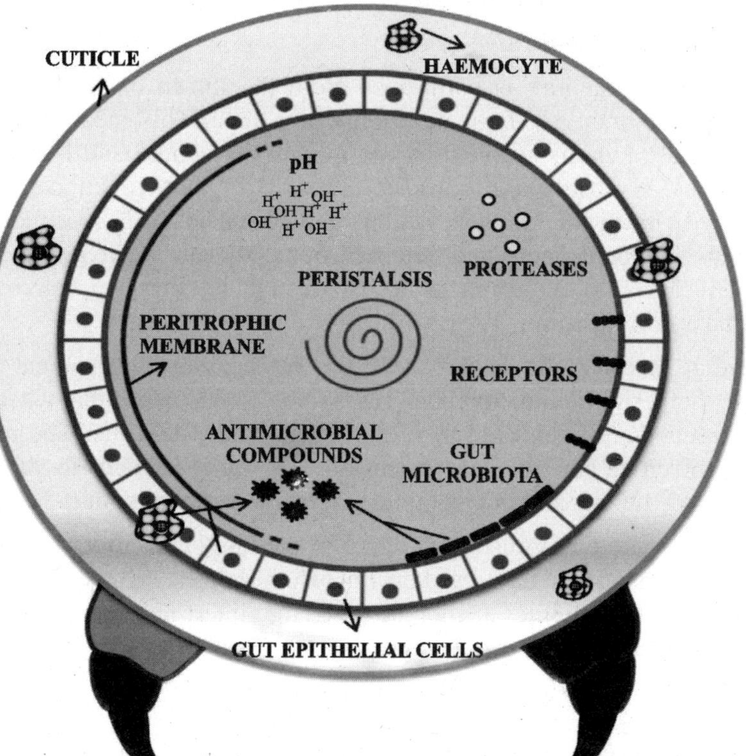

Fig. 3.4. Schematic diagram of local defence mechanisms of an insect model against pathogenic microbes.

Case study - The termite *R. flavipes*, a social insect that constructs underground galleries, coats its tunnels with regurgitants and fecal excretions. The gut microbiota deposited on the tunnel walls serve as a biological barrier against resident soil insect mycopathogens.

In addition to the gut microbiota, the ingested food may contain chemicals that suppress or enhances microbial development. Plants produces a variety of phytoalexins that provides an acquired resistance against many disease agents. The majority of the phytoalexins (isoflavanoids, flavanoids, sesquiterpenes, diterpenes, glycoalkaloids *etc.*) display broad-spectrum activity against fungi, bacteria, and parasites. These stress metabolites are also elicited by insect herbivory and are considered to be potent insect antifeedants. Certain plant chemicals may serve as precursors and require the insect to process them into defensive components.

Case study - Silkworm larvae, when fed fresh mulberry leaves, are rarely infected with enteric diseases. However, when these larvae are fed artificial diet they are often infected by the opportunistic intestinal bacterium *Streptococcus faecalus*. Mulberry leaves contain high levels of chlorogenic acid that is processed by the digestive enzymes into caffeic acid. Caffeic acid within the alkaline gut environment is converted to caffeoquinone, which complexes to amino acids, giving it potent anti-streptococcus activity.

3.4.2. Peritrophic membrane

The columnar epithelium is partitioned from the lumen or food bolus by the peritrophic membrane (PM) or peritrophic envelope. This membrane, found in several animal phyla, is secreted by the midgut epithelium and is comprised of glycoproteins, glycosaminoglycans *etc*. The PM is multi-functional: It protects the microvilli from the abrasive action of ingested material; partitions the gut lumen into endoperitrophic and ectoperitrophic spaces; prevents the uptake of allelochemicals; and it serves as a barrier against ingested microorganisms (Eisemann and Binnington, 1994).

Several insect pathogens possess the physical or biochemical means to penetrate the peritrophic membrane. The spores of microsporidia, when ingested by host insects, are stimulated to release their polar filaments. The mechanical forces involved in the filament extrusion are strong enough to penetrate the peritrophic membrane and in certain instances, the entire midgut tissue.

Case study 1 - Certain viruses belonging to the family Baculoviridae have been reported to contain factors in their inclusion bodies which are capable of degrading the peritrophic membrane and the intestinal mucins coating the microvillar surface.

Case study 2 - The ookinetes of the malarial parasite *Plasmodium gallinaceum* produce and secrete *chitinase*, which has been proposed to assist in their penetration through the PMs of the mosquito vector (Huber *et al.,* 1991).

3.4.3. Midgut epithelium

The columnar cells are the dominant cell type found in insect midgut tissue. Columnar cells, like the polarized vertebrate intestinal cells, are characterized by their striated border (brush border). The brush border, comprised of numerous microvilli, provides an extensive surface area required for nutrient digestion and assimilation. The insect microvillar surface, like the vertebrate intestinal layer, is coated with mucins. Unfortunately, very little is known about insect mucins.

The basal side of the midgut epithelium is covered by a well-defined basement membrane. This extracellular matrix has multifunctional roles: it compartmentalizes the insect tissues; serves as a support matrix for cells; acts as a molecular sieve; and regulates cell behaviour. The major components of the basal membrane are the exocellular matrix (ECM) proteins, including collagen, laminin, and various other proteoglycans and glycosaminoglycans.

In insects the ECMs comprising the basement membranes are synthesized mainly by haemocytes and fat body cells. Potentially, these structures, affiliated with either the midgut or other insect tissues, may serve as barriers to microbial entry. The midgut epithelium, although lacking the rigid cuticle barrier, may mediate an effective defence response against microbial invasion. The damage induced by pathogen ingress may stimulate columnar cells to detach from the basal membrane and to be sloughed off into the midgut lumen. This sloughing off of damaged midgut cells is reminiscent of the apoptosis response. Histological studies have demonstrated that this sloughing of damaged cells is one mechanism by which insects resist various cytoplasmic and nuclear polyhedrosis viruses

Histological studies have demonstrated that various insect viruses fuse to the microvillar membranes. Whether these particles are recognized by general or specific receptors is not yet resolved. Based on other virus-animal models, one would expect that the virus-cell receptor interaction dictates the specificity of certain insect viruses. The relative abundance and distribution of receptors have been shown to play key roles in the susceptibility of insects to bacterial toxins.

Case study 1 - Laboratory populations of *Culex quinquefasciatus* larvae resistant (10 fold) to *Bacillus sphaericus* did not possess the class-saturable specific receptors produced by susceptible larvae.

Case study 2 - Selected populations of *Heliothis virescens, Plodia interpunctella, Plutella xylostella,* and *Spodoptera exigua* selected for expressed resistance to the 6-endotoxin of *B. thuringiensis* possess a microvillar membrane with an altered and/or reduced receptor population.

3.5. Spread of Pathogens in the Host Insect

Insect pathogens, upon penetrating the cuticle and peritrophic membranes, invade the underlying epithelial tissues and either establish a localized infection or spread to secondary target tissues, causing a systemic infection. Unfortunately, very little is known about the dispersal of most insect pathogens within the host insect.

In general, pathogens that cause localized infections cause chronic diseases in insect hosts.

Case study 1- The Cypovirus, after being processed in the gut-lumen, invades the midgut columnar epithelial cells. Within infected cells, non-occluded virus progeny are produced in the cytoplasm and, although capable of infecting adjacent cells, are unable to breach the basement membrane to infect non-midgut tissues. At the late stage of infection, this virus produces many inclusion bodies containing large numbers of virus particles. The development of cypoviruses within the midgut stimulates the shedding of the infected cells into the lumen, resulting in the release of numerous occluded and non-occluded virus particles. These viruses may infect replacement columnar cells or be voided with fecal material.

Case study 2- Similarly, certain microsporidia, through extrusion of their polar filaments, infects midgut cells, where they undergo a cycle of intracellular replication. Infected cells containing progeny spores are sloughed off into the lumen. The spores activated by the gut conditions infect a second generation of host cells. The ability of the host insect to survive such infections depends upon its ability to regenerate midgut cells and to quickly void the inoculum released into the lumen. Whether the regenerative cells display phenotype plasticity and produce resistant replacement cells has yet to be investigated.

Unlike the cypoviruses, the baculoviruses are polytropic (which can affect more than one type of cell) and when ingested by host insects, are capable of infecting a wide array of host tissues of midgut cells. Within these cells, viral replication is usually limited and results in the production of progeny non-occluded viruses. The mechanism(s) by which these viruses breach the basement membrane is unclear. In the case of the nuclear polyhedrosis viruses (NPVs), it has been proposed that non-occluded viruses infect the tracheole cells that permeate the midgut tissue. Utilizing the tracheal system as a channel, progeny viruses are able to spread to all insect tissues.

Other baculoviruses, such as the granulosis viruses (GV), which do not replicate in the tracheal epithelium, are believed to breach the basal membranes, gaining access to susceptible fat body tissues.

In the case of the bacterium *Bacillus popilliae*, vegetative cells released from germinating spores in the lumen gain access to the columnar cells by a phagocytic process. Initial replication occurs within a phagocytic vacuole. Infection of the midgut results in the recruitment of haemocytes, which adhere to midgut tissues by an undefined mechanism, and the vegetative bacterial cells are translocated from the midgut to haemocytes. Within the haemocytes these cells undergo additional cycles of intracellular replication prior to release into the hemocoel.

The mechanisms by which this prokaryote survives the intracellular environment are similar to those strategies used by vertebrate pathogens. For example, baculoviruses are known to produce specific virus-encoded proteins that effectively turn off the apoptotic or host cell death program. If operational, this program, triggered by cell damage, results in cell suicide which terminates viral development (for review, see Teodoro and Branton, 1997). Certain pathogens, such as the insect pathogenic fungi, completely bypass the epithelial tissues and penetrate directly through the host cuticle and its underlying epidermis. Within the hemocoel specialized fungal cells are produced which are capable of extracellular replication. Similarly, other pathogens, such as the nematode-vectored bacteria, gain direct access to the host hemocoel where they multiply as extracellular pathogens.

3.6. Survival and Replication of Pathogens in the Host Insect

The ability of both intracellular and extracellular pathogens to multiply rapidly play a key role in the disease process. Many prokaryotes, when provided appropriate nutrients, will undergo *binary fission*, producing a set of daughter cells every 30–60 minutes. A single bacterium in the exponential growth phase produces 2^{24} daughter cells within 24 hours. Intracellular pathogens such as viruses undergo bursts of population growth. In cell culture, viruses may complete their developmental cycle within hours, producing and releasing hundreds to thousands of infectious particles.

Similarly, the extracellular mycopathogens have generation times that are only a fraction of those of their hosts. In general, the *in vivo* growth rates of most pathogens are slower than those calculated under *in vitro* conditions. The host system may restrict certain micro- or macronutrient pools, providing a sub-optimal growth environment for a pathogen. For example, vertebrates bind up nearly all of their elemental iron in *transferrin*-type molecules, making it unavailable to invasive microbes. In response to this iron-depleted environment, many intracellular and extracellular pathogens synthesize *siderophores*, scavenger molecules with a high affinity for iron for survival.

In other cases, the outer surface of insect pathogens may be covered with a material which is neutral to the host. For example, the cell wall of the hyphal body stage of the fungus *Nomuraea rileyi* is covered with an outer amorphous layer. This layer, synthesized by the fungus, possesses epitopes common to those found on the basement membrane of the host. This type of molecular mimicry is a strategy that provides pathogens, a means to evade the host non-self-recognition system.

In certain cases, masking with host components may assist in the horizontal dissemination of the disease agent. The virally modified host membrane of many budded viruses is often the critical determinant for virus uptake by susceptible cells. The *in vivo* pathogens may secrete metabolites which debilitate insect internal defence systems. The major targets that are normally suppressed are the plasmatocyte and granulocyte components of the insect cellular defence system.

The parasitoid-related poly DNA virus group is one of the best-studied examples of microbe-induced immunosuppression of insects. These viruses are delivered into host caterpillars during oviposition and invade the haemocytes. Poly DNA virus gene expression within these haemocytes results in the synthesis of proteins which paralyzes the ability of these cells to respond to non-self. The poly DNA virus-coded proteins in concert with various parasitoid-derived components modify the host to create an environment suitable for parasitoid growth.

Many other organisms, including the mycopathogens *Metarhizium anisopliae* and *Beauveria bassiana*, are known to secrete metabolites which debilitate (make week) insect cell defences. The fungal cyclic peptides *destruxins* and *beauvericin*, acting as channel-forming agents, inactivate certain haemocyte classes. In addition to evading or manipulating insect defences, certain pathogens are able to tolerate host antimicrobial response.

3.7. Mechanisms of Cell and Tissue Damage

For microbes to be defined as pathogenic, they must damage the host. The viruses directly damage the host cells/tissues. In several cases, virus infection leads to an immediate shutdown of cell transcription and translation. Additionally, many viruses alter the permeability of the cell membranes, causing cellular or nuclear hypertrophy. Hypertrophic conditions alter cell-to-cell contact and often result in the sloughing of infected cells from the tissue layer. The presence of intracellular pathogens often stimulate cell apoptosis, resulting in the fragmentation of the infected cells.

In general, intracellular pathogens requiring a functional cell for their survival do not produce or release powerful toxins. Many extracellular pathogens secrete biologically active metabolites that damage the host and assist in disease development. These metabolites, or toxins, include the enzymes that digest the basement and cell membranes, the various exotoxins, and the endotoxins which are components of the outer membrane of gram negative bacteria. The exotoxins, some of the most potent biomolecules, may act as pore-formers and disrupt membrane permeability and/or enter the cell and inhibit cell function. The major bacterial endotoxin, lipopolysaccharide (LPS), is known to activate the complement and coagulation cascades and stimulates the release of vasoactive substances. Overstimulation of the vertebrate immune system by lipopoly-saccharide (LPS) is harmful and may results in shock and death.

In insects, lipo-polysaccharide (LPS) is a major elicitor for the humoral and cellular defence systems, stimulating the production of a cascade of cationic peptides and/or the induction of the phenoloxidase (PO) cascade. In fact, lipo-polysaccharide (LPS) receptors have been isolated and characterized from several lepidopteran hosts. Activation of the phenoloxidase (PO) cascade may results in the uncontrolled production of melanin, resulting in death. In comparison to the vertebrate immune system, the insect defence activities are primitive. The classical immunopathology caused by the vertebrate antibody-antigen associations, *i.e.,* Anaphylactic, cell-mediated, cytotoxic, and immune-complex reactions, have not been observed in insects.

3.8. Pathogen Emergence and Transmission

The final stage in the infection is the production and release of progeny. Many of the well-studied disease agents of insects are extremely efficient in converting insect biomass into infectious progeny. These disease agents in an ecological framework would all be considered as *r-strategists*. Microscopic examination of insects at the late stage of disease often reveals a tremendous number of progeny in these hosts (Fig. 3.5).

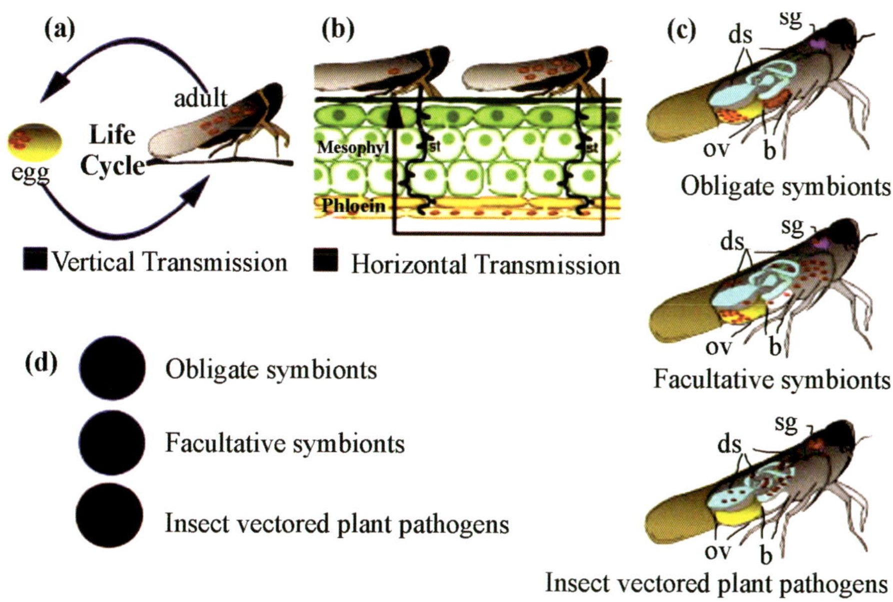

Fig. 3.5. General strategies of Transmission of insect pathogen (PC: Bressan, 2014)

Case study 1 – The haemolymph of larvae infected with *Nomuraea rileyi* may contain 10 fungal cells per micro liter of host haemolymph.

Case study 2 – Similarly, the haemolymph of scarab larvae infected with *Bacillus popillae* contain so many bacterial spores that the haemolymph becomes milky. Although completely filled with bacterial spores, these grubs may remain alive and active for weeks.

One of the most impressive examples of insect tissue-to-pathogen conversion is observed in insects infected with certain strains of baculo-viruses. By the late stage of infection, these viruses invade virtually all insect tissues and convert host cells into highly efficient viral production factories. These reprogrammed host cells produce hundreds of viral occlusions within their hypertrophied nuclei. At death, varieties of viral-encoded hydrolases are produced and function to disrupt the tissues and exoskeleton. Within hour's post-mortem, diseased larvae liquefy, releasing 10^7–10^8 virus occlusions into the environment. The intensity, speed, and complete destruction of tissues caused by Baculovirus replication have no parallel in vertebrate hosts.

A second example of effective insect tissue-to-pathogen conversion is observed with many mycopathogens. As mentioned earlier, prior to host death, these extracellular pathogens reach extremely high densities in the hemocoel. After attaining a critical density, these pathogens may switch from the vegetative growth phase to a tissue invasive phenotype. Within hours this mycelial form completely destroys host tissue and mummifies the host. At this stage, the vast majority of host tissues have been converted to fungal biomass. The well-known insect pathogens complete their *in vivo* development with the production of a resistant stage such as the virus occlusion, the protozoan cyst, or the fungal and bacterial spores. The specificity of these disease agents, the short-term availability of susceptible hosts, and the lack of vector involvement are some of the factors that dictate this life style.

In most cases, insect pathogens infect a limited number of insects and do not have reservoir hosts for disease maintenance. Unlike vertebrates, the insect is short-lived and is not normally available as a host on a year-round basis. Lastly, insect pathogens are generally dispersed passively and lack the vector component responsible for transmitting and maintaining many of the vertebrate disease agents. The packaging of pathogens into occlusions, spores, or cysts does provide the disease with two survival options-

1. Firstly, the resistant structure, after being released, can contact and infect a susceptible. For example, in the case of a multi-voltine host, susceptible are available for contact with the pathogens released from diseased insects. In many cases, activation of the resistant structures require access to host-derived signals. For example, the baculoviruses will release the occluded virus particles after being exposed to host digestive fluids. If ingested by non-hosts, such as a vertebrate or insect predator, intact occlusions are frequently passed through their digestive tract, resulting in the passive spread of the disease agents.
2. In the second scenario, the resistant structures do not contact a susceptible, but serve as the dormantor over wintering disease stage. Many of the structures released from diseased insects can survive for months or years in the environment, providing ample opportunity to eventually contact a susceptible

References

Andersen, S.O., M.G. Peter, and P. Roepstorff (1996). Cuticular sclerotization in insects. *Comp. Biochem. Physiol.* 113B: 689–705.

Ashida, M. and P.T. Brey (1995). Role of the integument in insect defense: Pro-phenol oxidase cascade in the cuticular matrix. *PNAS*. 92:10698–10702.

Bailey, L. (1981). Honeybee Pathology. Academic Press. NY. pp 124.

Boucias, D.G., C. Stokes, G. Storey and J. Pendland (1996). Effect of imidacloprid on both the termite, *Reticulitermes flavipes* and its interaction with insect pathogens. *Pfanzenshutz Natrichten Bayer* 49: 103–144.

Bressan, A. (2014). Emergence and evolution of Arsenophonus bacteria as insect-vectored plant pathogens. *Infection, Genetics and Evolution, 22:* 81–90.

Brey, P.T., Lee, W.J., Yamakawa, M., Koizumi, Y., Perrot, S., Francois, M. and Ashida, M. (1993). Role of the integument in insect immunity: epicuticular abrasion and induction of cecropin synthesis in cuticular epithelial cells. *Proceedings of the National Academy of Sciences, 90*(13): 6275–6279.

Buckner, J.S., M.C. Mardaus, and D.R. Nelson (1996). Cuticular lipid composition of *Heliothis virescens* and *Helicoverpa zea* pupae. *Comp. Biochem. Physiol.* 114B: 207–216.

Dillon, R.J. and C. Charnley (1991). The fate of fungal spores in the insect gut. In: The Fungal Spore and Disease Initiation in Plants and Animals (Ed. G.T. Cole and H.C. Hoch). *Plenum Press.* NY. pp 129–156.

Drion, B. and Pendland, J.C. (1998). *Principles of Insect Pathology* (No. 595.7 B66).

Eisemann, C.H. and K.C. Binnington (1994). The peritrophic membrane: its formation, structure, chemical composition and permeability in relation to vaccination against ectoparasitic arthropods. *International J. Parasitology.* 24(1): 15–26.

Federici, B.A., Park, H.W., Bideshi, D.K., Wirth, M.C. and Johnson, J.J. (2003). Recombinant bacteria for mosquito control. *J. Exp. Biol.* 206: 3877–3885.

Glancy, B.M., R.K. VanderMeer, A. Glover, C.S. Lofgren and S.B. Vinson (1981). Filtration of microparticles from liquids ingested by the red imported fire and Solenopsis invicta Buren. *Insectes Sociaux.* 28: 395–401.

Huber, M., Cabib, E. and Miller, L.H. (1991). Malaria parasite chitinase and penetration of the mosquito peritrophic membrane. *Proceedings of the National Academy of Sciences, 88*(7): 2807–2810.

Hurst, M.R., (2016). Non-spore-Forming Bacterial Entomopathogens: Their Toxins, Hosts and the Environment: Why Be a Pathogen. In *The Rasputin Effect: When Commensals and Symbionts Become Parasitic* (pp. 169–220). Springer, Cham.

Ortiz-Urquiza, A. and Keyhani, N.O. (2013). Action on the surface: entomopathogenic fungi versus the insect cuticle. *Insects, 4*(3): 357–374.

Raes, H. and Verbeke, M. (1994). Light and electron microscopical study of two types of endocrines cell in the midgut of the adult worker honeybee (*Apis mellifera*). *Tissue and Cell, 26*(2): 223–230.

Terra, W.R. and Ferreira, C. (2012). Biochemistry and molecular biology of digestion. In *Insect molecular biology and biochemistry* (pp. 365–418).

Teodoro, J.G. and Branton, P.E. (1997). Regulation of apoptosis by viral gene products. *Journal of virology, 71*(3): 1739.

Vallet-Gely, I., Lemaitre, B. and Boccard, F. (2008). Bacterial strategies to overcome insect defences. *Nature Reviews Microbiology, 6*(4): 302.

Willis, J.H. (1996). Metamorphosis of cuticle, its proteins, and their genes. In: *Metamorphosis.* Academic Press, NY. pp 253–282.

Question Time

1. Describe various stages of infection process into host by pathogen in brief.

2. Illustrate factors affecting pathogens establishment in host body with appropriate diagram.
3. Discuss the mechanisms of cell and tissue damage by pathogen with suitable examples.

True or False

1. The infection process refers to the germination of an infective propagule on the surface of a potential insect host – **False**
2. Insect pathogens may gain entry into insects only through mouth and anal routes – **False**
3. Certain fungi and nematodes can penetrate through the respiratory system of an arthropods – **True**
4. The physical and chemical properties of the insect cuticle produce a highly effective barrier against microbes – **True**
5. The mesenteron (midgut) region, like the stomodeum (foregut) and proctodeum (hindgut), is lined with cuticle – **False**
6. Digestive enzymes can help in establishment of the pathogen in insect gut – **True**
7. For microbes to be defined as pathogenic, they must damage the host – **True**
8. *Metarhizium anisopliae* and *Beauveria bassiana*, are known to secrete metabolites which debilitate (make week) insect cell defences – **True**

■■■■

Chapter 4

CONCEPT OF EPIZOOTIOLOGY

Epizootiology is a central component of insect pathology that affects diverse aspects of the discipline. Epizootiology is defined as the study of animal disease dynamics on the basis of mass phenomena (Steinhaus, 1967; Fuxa and Tanada, 1987; Onstad *et al.,* 2006). Thus, epizootiology deals with disease on a population level rather than an individual basis. The subject concerns both infectious and non-infectious diseases (though this chapter will focus primarily on the former) and affects all pathogen groups. Epizootiology deals with epizootic and enzootic levels of animal disease. An epizootic is defined as an outbreak of disease in which there is an unusually large number of cases, whereas an enzootic refers to a low level of disease that is constantly present in a population (Steinhaus, 1967; Onstad *et al.,* 2006). Symptomatology and Etiology are the two branches of science which help us in understanding epizootiology profoundly. Symptomatology is the branch of science concerned with the study and classification of the symptoms of disease. Whereas, the word etiology comes from the Greek *etio-*, which means 'causation' and *-ology*, which refers to the scientific study of something. Hence, etiology is the branch of science that studies the causes of diseases and the factors underlying their spread.

The study of epizootiology crosses multiple interests within insect pathology. Epizootiology is the foundation for studying disease fluctuation in natural insect populations from an ecological perspective. However, the primary motivation of many insect pathologists is not the study of natural disease levels, but the use of diseases to suppress insect pest populations, *i.e. Microbial control.* Microbial control can be viewed as applied epizootiology with the goal of inducing an epizootic in the targeted insect population through manipulation. In contrast, many insect pathologists are motivated by reducing or eliminating the prevalence of disease in insect populations, *e.g.* in *beneficial insects (e.g. pollinators*) or insects in insectary. In either case, understanding factors that cause epizootics is critical to the successful implementation of microbial control and equally important in developing methods to guard against epizootics.

Epizootiology: Basic Principles

A clear understanding of the basic principles of epizootiology is necessary to understand fully the science of causes and forms of disease. Given that the basic definition of an epizootic entails an unusually large number of cases of a disease in a host population, the question arises: what constitutes an unusually large number?

To answer this question, one must establish what the long-term prevalence of the disease is in a space-time framework. Prevalence is defined as the total number of cases of a particular disease at a given time, in a given population (Onstad *et al.*, 2006). The prevalence of a causal agent at any point in time from any location can then be subjected to statistical analysis to establish when prevalence is outside the norm. While epizootics are sporadic in their occurrence and marked by a sudden change in prevalence, enzootics are present over very long periods and their prevalence in the host population varies little.

The four basic components of an epizootic are :

- The pathogen population,
- The host population,
- Transmission, and
- The environment.

These components work alone or in combination in a manner that is conducive for the development of an epizootic or precludes an epizootic from taking place. In the field, it is the sum effect of all of these factors that determines whether or not an epizootic takes place.

The Pathogen Population

Numerous properties associated with the pathogen population (*e.g.*, pathogen density, dispersal, infectivity and latency, virulence, and genetics) play a large role in determining whether or not conditions are in place for an epizootic wave to be initiated.

- Foremost, the pathogen **population density** must be sufficient to initiate an epizootic.
- In addition, within the vicinity of the host, the pathogen population must also be able to **disperse** sufficiently, actively or passively, to reach the host, and once making contact, the pathogen must be able to infect; therefore, dispersal and infectivity are among the key factors of importance.
- Once reaching the host, the pathogen's **level of virulence** is of utmost importance.
- Furthermore, there are various genetic factors associated with pathogen strains that may influence **pathogenicity and virulence**.
- Each of the properties associated with the pathogen population alone or in combination with one or more of the others may have a large impact on whether or not disease occurs within a host and subsequently spreads to other individuals in the population resulting in an epizootic.

These pathogen properties, which vary widely among pathogen groups, are discussed below:

A. **Pathogen density**: Pathogen density includes the spatial distribution of the pathogen (D'Amico *et al.,* 1996). Although it is a basic character of a pathogen, it depends on other two characteristics (*e.g.* pathogen reproductive rate and capacity to survive). A pathogen with a high density and widespread spatial distribution in the field is inherently well suited to cause an epizootic. However, pathogens with low densities in the field or limited spatial distribution can result in localized enzootics, and under favourable biological and environmental conditions can result in epizootics (Dwyer and Elkinton, 1993).

The implementation of a microbial control program most often attempts to flood an entire crop with high levels of a pathogen and initiate an epizootic. Timing of the pathogen's application in microbial control programs is critical so that the pathogen and the susceptible stage of the insect occur simultaneously in the environment.

B. **Dispersal:** The ability of a pathogen to disperse, either under its own power or passively, has implications on pathogen spatial distribution. Most pathogens have limited or no capacity to disperse on their own. Entomopathogenic nematodes in the families, Steinernematidae and Heterorhabditidae are the best example of terrestrial Entomopathogens capable of dispersing under their own power. Other pathogens that use the soil as a reservoir (viruses, fungi, bacteria) have no capacity to disperse on their own. Rather, these pathogens rely on biotic and abiotic factors to distribute them in the environment.

C. **Infectivity and Latency:** Infectivity is closely related to transmission, as it is the method of entry into a potential host. Routes of entry include oral ingestion and entry via the digestive tract, direct cuticular penetration, tracheal entry, entry via the reproduction system, or entry via the action of parasitoids. Pathogens with multiple routes of entry into their host (*for example-* Fungi and Nematodes) are expected to result more easily in epizootics as disease transmission is facilitated (Tanada and Fuxa, 1987).

Some pathogen infections may be latent (*i.e.* Non-infective and non-replicative state) until triggered by some environmental pressure. However, the development of a visible infection after an environmental pressure is not proof that a disease is latent, as would be the case for pathogens causing chronic infections (Tanada and Fuxa, 1987).

Case study - Overwintering European corn borer (*Ostrinia nubilalis*) larvae also harbour latent *Beauveria bassiana* infections that become active with the breaking of larval diapauses, resulting in fungal spores being produced and released into the environment in spring when insect hosts are abundant (Bruck and Lewis, 1999) (Fig. 4.1).

Fig. 4.1. European corn borer *Ostrinia nubilalis* infected with biological control fungus, *Beauveria bassiana,* 7 to 10 days (A); 10–15 days (B); at completion of life cycle (C)

D. **Virulence:** The median lethal dose of a pathogen needed to kill 50 per cent of the tested insects (LD50) is a typical measure of virulence (Shapiro-Ilan *et al.,* 2005a). A more virulent pathogen requires fewer infective propagules to cause disease relative to a pathogen that is less virulent. Pathogen virulence can be manipulated in the laboratory by a variety of approaches such as: (1) repeated passages through susceptible hosts; (2) identification of strains within a species of pathogens with differential virulence; and (3) introducing pathogens with other substances to enhance their invasive ability or virulence.

The Host Population

The susceptibility of a host population to pathogen infection plays a critical role in determining whether or not an epizootic takes place. Inherent susceptibility of an insect to infection by a pathogen is genetically based. The more susceptible a host is to infection, the lower the pathogen dose necessary. The level of host susceptibility is most commonly measured with an estimate of LD50. The lower the LD50, the fewer propagules are necessary to cause mortality.

Epizootics generally occur at high host densities (Onstad and Carruthers, 1990). High host density facilitates transmission of pathogens between infected and healthy hosts and results in large numbers of infective propagules being produced and released into the environment. Yet, epizootics can occur in situations where pathogens are widely distributed and host populations are low (Dwyer and Elkinton, 1993). Certain properties like **Genetic Resistance**, Behavioural **Resistance** and **Host Associations** can affect susceptibility of a host population to pathogen infection. We will discuss it in brief:

A. **Genetic Resistance:** Host resistance is one mechanism that can cause a substantial reduction in host susceptibility.

Case study 1 - The development of resistance to a pathogen outside the laboratory occurs in the Indian meal moth, *Plodia interpunctella*, exhibiting resistance to *Bacillus thuringiensis* (McGaughey, 1985). Moths isolated from treated grain bins were more resistant than those isolated from untreated bins, indicating that resistance developed quickly in the field.

Case study 2 - The first reported resistance to *Bacillus thuringiensis* sprays in the field was observed in *Plutella xylostella*, as the large-scale planting of crops engineered to produce the protein toxins (Bt. crops) represents intense selection pressure for development of resistance to the gene product of an insect pathogen. (Tabashnik *et al.*, 1990).

B. Behavioural Resistance: Hosts exhibit a wide range of behavioural responses to the presence of pathogen propagules in the environment before infection, in addition to behavioural responses after infection. The type of behaviour response can vary widely. Many responses help the host either to avoid or to mitigate infection, while others aid the pathogen in subsequent dissemination of propagules in the environment.

Case study 1 - Termites reared in groups physically removed 80 per cent of *Metarhizium anisopliae* conidia (Fig. 4.2) from their nest mates and eliminated the conidia through the alimentary tract, while individually reared termites did not reduce their surface contamination (Yanagawa and Shimizu, 2007).

Fig. 4.2. *Metarhizium anisopliae* completing its life cycle, by producing conidia on a dead *Coptotermes formosanus* soldier. (Termite) Scale bar = 1 (source mm.doi:10.1371/ journal.pone.0034484.g001 https://www.researchgate.net/publication/ 223983163_ When_subterranean_termites_challenge_the_rules_of_fungal_epizootics).

Case study 2 - Red imported fire ants, *Solenopsis invicta*, bury nest mates infected with *Beauveria bassiana* to reduce transmission (Pereira and Stimac, 1992).

Case study 3 -Mole crickets modify their behaviour in response to *Metarhizium anisopliae* and *Beauveria bassiana* in a way that reduces their exposure to these fungi (Villani *et al.*, 2002; Thompson and Brandenburg, 2005).

C. Host Associations: Pathogens have different capacities for surviving in primary or secondary hosts. As such, pathogens capable of infecting more than one host in the environment are often better able to survive and persist (Goettel, 1995).

Primary pathogen survival and replication in the environment take place on primary hosts. But pathogens with long periods of infection that do not kill the host quickly are better suited to persist in the environment as they do not destroy the host population too quickly.

Case study- Infection of the chronic pathogen *Nosema pyrausta* in *Ostrinia nubilalis* populations is cyclic, with alternating enzootic and epizootic patterns over a period of several decades (Lewis *et al.,* 2009).

Secondary hosts (also called intermediate or alternate hosts) are host species in which immature, intermediate, or asexual stages of a pathogen occur (Onstad *et al.,* 2006).

Case study - An early example of an Entomopathogens using an obligatory secondary host for completion of its life cycle involves the transmission of the microsporidium *Amblyospora* sp. infecting mosquitoes to alternate copepod (a minute crustacean) hosts.

The ability to infect a wide range of hosts beyond the primary target, none of which is obligatory, provides additional resources for pathogen survival.

Case study - *Beauveria bassiana* is a well-studied example of a facultative pathogen with a broad host range, infecting over 700 species of arthropods from different orders (Li, 1988; Meyling *et al.,* 2009).

Transmission

Transmission is the process by which a pathogen or parasite is passed from a source of infection to a new host (Onstad and Carruthers, 1990). Insect pathogens have evolved a wide array of mechanisms to ensure their long-term survival and transmission to new hosts. Insect pathogens are primarily transmitted either directly host to host, or from the infected host into the environment, where they are subsequently acquired by a susceptible host.

Methods of Transmission and Modes of Dissemination are two defining characters under this component.

A. **Methods of Transmission:** Transmission can be vertical or horizontal. Horizontal transmission is the transfer of a pathogen from individual to individual, either host to host, or host to environment to host (Canning, 1982). Vertical transmission, also referred to as congenital, parental, or hereditary, is the direct transmission of a pathogen from parent to progeny (Andreadis, 1987)

Case study 1 - Host to-host horizontal transmission of *Beauveria bassiana* and *Metarhizium anisopliae* between tsetse flies, *Glossina morsitans* (Kaaya and Okech, 1990), the Mexican fruit fly, *Anastrepha ludens* (Toledo *et al.,* 2007), and the German cockroach, *Blatella germanica* (Quesada-Moraga *et al.,* 2004) has been well demonstrated.

Case study 2 - Horizontal transmission of bacterial and viral pathogens primarily occurs via the host-to-environment-to-host pathway. However, direct horizontal transfer from host to host does occur in the case of cannibalism in which an insect consumes infected conspecifics (Chapman *et al.*, 1999).

Many insect viruses are vertically transmitted in the environment. Vertical transmission has been best studied in baculoviruses, but even within this group, the nature and role of vertical transmission on epizootiology are poorly understood (Cory and Meyers, 2003).

Case study - There is strong evidence of vertical transmission of NPV (Olofson, 1989) and cytoplasmic polyhedrosis viruses (also known as cypoviruses) of Lepidoptera (Sikorowski *et al.,* 1973), indicating continuation of gut infections initiated in the larval stage. In these cases, transovum transmission as an external contaminant of the deposited eggs appears to be the most likely route of entry.

B. **Modes of Dissemination:** Dissemination is the ability of the pathogen to spread within a host population and throughout the environment (Tanada, 1963). There are four primary routes of pathogen dissemination in the field (Andreadis, 1987), based on:

   a. **The pathogen's motility**: Pathogen motility in general is very limited in terms of both the ability of pathogens to move under their own power and the distance they are able to travel in the environment

   b. **The behaviour and movements of the primary host**: Movement and activities of the primary host in the environment are important factors in pathogen dissemination.

- The direct mechanisms of pathogen dissemination (host to host) by the primary host include vertical transmission, cannibalism, and grooming.
- Indirect pathogen dissemination (host to environment to host) occurs via horizontal transmission by excretion of infective stages in feces and anal discharge (Fuxa *et al.*, 1998) and meconial discharges.
- Disintegration of cadavers with infective pathogen propagules (Fuxa, 2004), and elimination of infected exuvia during moulting.
- Perhaps the well-cited example of primary host behaviour that aids in the dissemination of disease is the treetop disease (also known as summit disease) of many lepidopterans and hymenopterans larvae.

Case study – The infected larvae migrate to the tops of the plant or tree on which they are feeding and subsequently die attached to leaves by their prolegs (Fig. 4.3). Being positioned in this way ensures that the foliage below is showered with inoculum as the cadaver breaks down. This facilitates pathogen transmission to healthy larvae on the plant (Fuxa, 2004).

   c. **The behaviour and movement of secondary hosts and non-host carriers:** Wind and rain are the most important abiotic environmental factors responsible for pathogen dispersal.

Fig. 4.3. This is a dead caterpillar killed by an infectious virus disease (Polyhedrosis), as identified by UC Davis butterfly expert Art Shapiro. (Photo by Kathy Keatley Garvey)

Case study 1 - Rainfall is an efficient means of transfer from the fungal entomopathogen reservoir in the soil to the surface of whorl-stage corn plants (Bruck and Lewis, 2002b). Plants receiving rainfall had a mean of 8.8 colony-forming units (cfu) of *Beauveria bassiana* per plant, while those not receiving rainfall had a mean of 0.03 cfu per plant.

Case study 2 - Prolonged rainfall can also transfer viral occlusion bodies from soil to the plant surface via rain splash (Fuxa and Richter, 2001).

Case study 3 - Wind was suspected in the dissemination up to 30 m of NPV (Hukuhara, 1973).

d. The actions of physical environmental factors: Epizootics of insect pathogens are heavily influenced by environmental factors (biotic and abiotic) and in many cases environmental factors are the most relevant in the epizootiological process (Ignoffo, 1992). The three primary factors contributing to the epizootics of disease the host population, the pathogen population, and transmission have already been discussed. Environmental factors may have direct or indirect impacts on any or all of these primary factors in the environment in which the host and pathogen occur (Benz, 1987).

Case study 1 - Conidial exposure of multiple isolates of *Metarhizium* sp. and *Beauveria* sp. indicates that the viability of all isolates drops remarkably with increasing exposure to UV radiation (Braga *et al.,* 2001).

Case study 2 - Temperatures below 15°C significantly slowed mycelial growth of *Metarhizium anisopliae in vitro* and *Otiorhynchus sulcatus* (commonly known as the black vine weevil) larval infection in soilless potting media (Bruck, 2007) (Fig. 4.4 and 4.5).

Case study 3 - The leaf collar of corn plants is a favourable habitat for *Beauveria bassiana* and *Bacillus thuringiensis,* providing an environment with adequate moisture and protection from UV radiation (McGuire *et al.,* 1994; Bruck and Lewis, 2002c; Lewis *et al.,* 2002).

Insect pathogens have evolved a variety of adaptations to overcome environmental barriers. By doing so, these fungi persist within the plant protected from sunlight.

Case study 1- There is clear evidence that the presence of an occlusion body increases virus survival in the field over free virions in the environment (Wood *et al.*, 1994; Cory and Evans, 2007).

Case study 2- Various fungal Entomopathogens have evolved the ability to grow endophytically with a wide variety of plants (Vega, 2008).

Fig. 4.4. *Otiorhynchus sulcatus* and *Beauveria bassiana* (PC: www.hs-geisenheim.de)

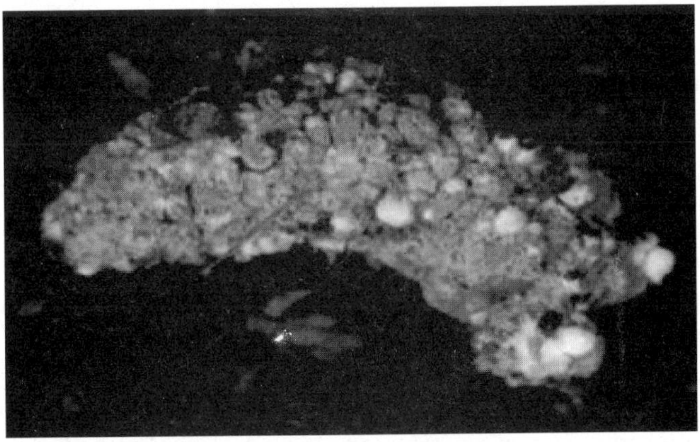

Fig. 4.5. An *Otiorhynchus sulcatus* larva engulfed by *Metarhizium* sp. (Image: Sara Reardon)

Future Thrust in Epizootiology

Although substantial progress has been made in the discipline of epizootiology, additional advances and broader integration of epizootiology into other components of invertebrate pathology are needed. The greatest barrier to expanded implementation of concept of epizootiology in microbial control is *lack of economic feasibility*. Thus, research is required to facilitate the use of microbial control agents in a cost-effective manner.

A larger emphasis on conservation approaches can augment microbial control usage without necessarily increasing the costs of crop production. In addition, the integration of microbial control tactics into more holistic, IPM programs may allow for improved cost competitiveness.

Advances in applied microbial control can be made through greater linkage with basic research. Fundamental research in insect pathology, such as on host-pathogen relationships at the molecular and organismal level, ecological relationships, physiology etc., is the foundation for microbial control and serves as a basis for the future growth and development of the discipline.

References

Andreadis, T.G. (1987). Transmission. In J.R. Fuxa & Y. Tanada (Eds.), *Epizootiology of Insect Diseases* (pp. 159–176). New York: Wiley Interscience.

Benz, G. (1987). Environment. In J.R. Fuxa and Y. Tanada (Eds.), *Epizootiology of Insect Diseases* (pp. 177–199). New York: Wiley Interscience.

Braga, G.U., Flint, S.D., Messias, C.L., Anderson, A.J. and Roberts, D.W. (2001). Effects of UV-B on conidia and gerlings of the entomopathogenic hyphomycete Metarhizium anisopliae. Mycological Research 105(7): 874–882.

Bruck, D.J. (2007). Efficacy of *Metarhizium anisopliae* as a curative application for black vine weevil (*Otiorhynchus sulcatus*) infesting container-grown nursery crops. *J. Environ. Hort.*, 25: 150–156.

Bruck, D.J. and Lewis, L.C. (1999). *Ostrinia nubilalis* (Lepidoptera: Pyralidae) larval parasitism and infection with entomopathogens in cornfields with different border vegetation. *J. Agric. Urban Entomol.,* 16: 255–272.

Bruck, D.J. and Lewis, L.C. (2002b). Rainfall and crop residue effects on soil dispersion and *Beauveria bassiana* spread to corn. *Appl. Soil. Ecol.*, 20: 183–190.

Bruck, D.J. and Lewis, L.C. (2002c). Whorl and pollen-shed stage application of *Beauveria bassiana* for suppression of adult western corn rootworm. *Entomol. Exp. Appl.*, 103: 161–169.

Canning, E.U. (1982). An evaluation of protozoal characteristics in relation to biological control of pests. *Parasitology,* 84: 119–149.

Chapman, J.W., Williams, T., Escribiano, A., Caballero, P., Cave, R.D. and Goulson, D. (1999). Age-related cannibalism and horizontal transmission of a nuclear polyhedrosis virus in larval *Spodoptera frugiperda*. *Ecol. Entomol.*, 24: 268–275.

Cory, J.S. and Evans, H.F. (2007). Viruses. In L.A. Lacey & H.K. Kaya (Eds.), *Field Manual of Techniques in Invertebrate Pathology* (2nd ed.). (pp. 149–174) Dordrecht: Springer.

Cory, J.S. and Myers, J.H. (2003). The ecology and evolution of insect baculoviruses. *Annu. Rev. Ecol. Evol. Syst.,* 34: 239–272.

D'Amico, V., Elkinton, J.S., Dwyer, G., Burand, J.P. and Buonaccorsi, J.P. (1996). Virus transmission in gypsy moths is not a simple mass action process. *Ecology*, 77: 201–206.

Dwyer, G. and Elkinton, J.S. (1993). Using simple models to predict virus epizootics in gypsy moth populations. *J. Anim. Ecol.*, 62: 1–11.

Fuxa, J.A., Fuxa, J.R. and Richter, A.R. (1998). Host-insect survival time and disintegration in relation to population density and dispersion of recombinant and wild-type nucleopolyhedroviruses. *Biol. Control*, 12: 143–150.

Fuxa, J.R. (2004). Ecology of insect nucleopolyhedroviruses. *Agric. Ecosyst. Environ.*, 103: 27–43.

Fuxa, J.R. and Richter, A.R. (2001). Quantification of soil-to-plant transport of recombinant nucleopolyhedrovirus: effects of soil type and moisture, air currents, and precipitation. *Appl. Environ. Microbiol.,* 67: 5166–5170.

Fuxa, J.R. and Tanada, Y. (1987). *Epizootiology of Insect Diseases.* New York: Wiley-Interscience.

Goettel, M.S. (1995). The utility of bioassays in the risk assessment of entomopathogenic fungi. In Biotechnology Risk Assessment: USEPA/USDA/Environment Canada/ Agriculture and Agri-Food Canada. Risk Assessment Methodologies. Proceedings of the Biotechnology Risk Assessment Symposium, June 6–8, 1995. Pensacola, FL (pp. 2–8). College Park: University of Maryland Biotechnology

Hukuhara, T. (1973). Further studies on the distribution of a nuclear polyhedrosis virus of the fall webworm, *Hyphantria cunea*, in soil. *J. Invertebr. Pathol.*, 22: 345–350.

Ignoffo, C.M. (1992). Environmental factors affecting persistence of entomopathogens. *Fla. Entomol.*, 75: 516–525.

Kaaya, G.P. and Okech, L.A. (1990). Horizontal transmission of mycotic infection in adult tsetse flies. *Glossina morsitans. Entomophaga*, 35: 589–600.

Keith Weller, USDA Agricultural Research Service, Bugwood.org

Lewis, L.C., Bruck, D.J. and Gunnarson, R.D. (2002). Measures of *Bacillus thuringiensis* persistence in the corn whorl. *J. Invertebr. Pathol.*, 80: 69–71.

Lewis, L.C., Bruck, D.J., Prasifka, J.R. and Raun, E.S. (2009). *Nosema pyrausta*: its biology, history, and potential role in landscape of transgenic insecticidal crops. *Biol. Control,* 48: 223–231.

Li, Z. (1988). In Y. Li, Z. Li, Z. Liang, J. Wu, Z. Wu and Q. Xu (eds.), Study and Application of Entomogenous Fungi in China. List on the insect hosts of *Beauveria bassiana*, Vol. 1 (pp. 241–255). Beijing: Academic Periodical Press.

McGaughey, W.H. (1985). Insect resistance to the biological insecticide *Bacillus thuringiensis. Science*, 229: 193–195.

McGuire, M.R., Shasha, B.S., Lewis, L.C. and Nelson, T.C. (1994). Residual activity of granular starch-encapsulated *Bacillus thuringiensis. J. Econ. Entomol.*, 87: 631–637.

Meyling, N.V., Lubeck, M., Buckley, E.P., Eilenberg, J. and Rehner, S.A. (2009). Community composition, host range and genetic structure of the fungal entomopathogen *Beauveria* in adjoining agricultural and semi natural habitats. *Mol. Ecol.*, 18: 1282–1293.

Olofsson, E. (1989). Transmission of the nuclear polyhedrosis virus of the European pine sawfly from adult to offspring. *J. Invertebr. Pathol.*, 54: 322–330.

Onstad, D.W. and Carruthers, R.I. (1990). Epizootiological models of insect diseases. *Annu. Rev. Entomol.,* 35: 399–419.

Onstad, D.W., Fuxa, J.R., Humber, R.A., Oestergaard, J., Shapiroilan, D.I., Gouli, V.V., Anderson, R.S., Andreadis, T.G. and Lacey, L.A. (2006). An Abridged Glossary of Terms Used in Invertebrate Pathology. http://www.sipweb.org/glossary. Society for Invertebrate Pathology.

Pereira, R.M. and Stimac, J.L. (1992). Transmission of *Beauveria bassiana* within nests of *Solenopsis invicta* (Hymenoptera, Formicidae) in the laboratory. *Environ. Entomol.,* 21: 1427–1432.

Quesada-Moraga, E., Santos-Quiro´s, R., Valverde-Garcı´a, P. and Santiago Alvarez, C. (2004). Virulence, horizontal transmission, and sublethal reproductive effects of *Metarhizium anisopliae* (anamorphic fungi) on the German cockroach (Blattodea: Blattellidae). *J. Invertebr. Pathol.*, 87: 51–58.

Reardon S. (2012). Murderous fungi feed their insect victims to plants. DOI: 10.1126/science.1222289

Shapiro-Ilan, D., Fuxa, J.R., Lacey, L.A., Onstad, D.W. and Kaya, H.K. (2005a). Definitions of pathogenicity and virulence in invertebrate pathology. *J. Invertebr. Pathol.*, 88: 1–7.

Sikorowski, P.P., Andrews, G.L. and Broome, J.R. (1973). Transovum transmission of a cytoplasmic polyhedrosis virus of *Heliothis virescens* (Lepidoptera: Noctuidae). *J. Invertebr. Pathol.,* 21: 41–45.

Steinhaus, E.A. (1967). *Principles of Insect Pathology* (2nd Ed.). New York: Hafner.

Tabashnik, B.E., Cushing, N.L., Finson, N. and Johnson, M.W. (1990). Field development of resistance to *Bacillus thuringiensis* in diamondback moth (Lepidoptera: Plutellidae). *J. Econ. Entomol.*, 83: 1671–1676.

Tanada, Y. (1963). In E.A. Steinhaus (Ed.), *Insect Pathology: An Advanced Treatise. Epizootiology of infectious diseases,* Vol. 2. New York: Academic Press.

Thompson, S.R. and Brandenburg, R.L. (2005). Tunneling responses of mole crickets (Orthoptera: Gryllotalpidae) to the entomopathogenic fungus, *Beauveria bassiana. Environ. Entomol.,* 34: 140–147.

Toledo, J., Campos, S.E., Flores, S., Liedo, P., Barrera, J.F., Villasenor, A. and Montoya, P. (2007). Horizontal transmission of *Beauveria bassiana* in *Anastrepha ludens* (Diptera: Tephritidae) under laboratory and field cage conditions. *J. Econ. Entomol.*, 100: 291–297.

Vega, F.E., Posada, F., Aime, M.C., Pava-Ripoll, M., Infante, F. and Rehner, S.A. (2008). Entomopathogenic fungal endophytes. *Biol. Control*, 46: 72–82.

Villani, M. G., Allee, L.L., Preston-Wilsey, L., Consolie, Y.X. and Brandenburg, R.L. (2002). Use of radiography and tunnel castings for observing mole cricket (Orthoptera: Gryllotalpidae) behavior in soil. *Am. Entomol.*, 48: 42–50.

Wood, H.A., Hughs, P.R. and Shelton, A. (1994). Field studies of the coocclusion strategy with a genetically altered isolate of the *Autographa californica* nuclear polyhedrosis virus. *Environ. Entomol.*, 23: 211–219.

www.hs-geisenheim.de/en/research/departments/crop-protection/professorship-for-crop-protection-in-viticulture-horticulture.

Yanagawa, A. and Shimizu, S. (2007). Resistance of the termite, *Coptotermes formosanus* Shiraki to *Metarhizium anisopliae* due to grooming. *Bio Control,* 52: 75–85.

Question Time

1. Describe Epizootiology in brief.
2. What are the basic components of an epizootic?
3. What are the properties associated with the pathogen population? Illustrate in detail.
4. How does mode of dissemination of a pathogen contribute in epizootic?
5. What are the routes of pathogen dissemination in the field?

One Word Answer

1. The first reported resistance to *Bacillus thuringiensis* sprays in the field was observed in – *Plutella xylostella.*
2. The process by which a pathogen or parasite is passed from a source of infection to a new host – Transmission.
3. The direct transmission of a pathogen from parent to progeny also referred to as congenital or parental transmission, is called – Hereditary transmission.
4. The ability of the pathogen to spread within a host population and throughout the environment – Dissemination
5. The method of entry of a pathogen into a potential host – Infectivity.

True or False

1. One of the basic components of an epizootic is soil texture. – **False**
2. A pathogen with a high density and widespread spatial distribution in the field is inherently well suited to cause an epizootic. – **True**
3. Pathogen density includes the spatial distribution of the pathogen – **True**
4. Pathogenicity and virulence are not affected by genetic factors associated with pathogen strains. – **False**
5. Pathogen application in microbial control program can be done round the year – **False**
6. LD_{50} is a typical measure of virulence of a pathogen – **True**
7. The susceptibility of a host population to a pathogen plays a crucial role in epizootic – **True**
8. UV radiation, temperature and moisture etc are physical and environmental factors which affect epizootical process – **True**

Chapter 5

DEFENCE MECHANISMS IN INSECTS AGAINST PATHOGENS

Executive Summary

For a pathogen to mount a successful infection on an insect, it must overcome three strategic components of the insect antipathogen defences: avoid recognition, overcome barriers, and withstand the innate immune response. The antipathogen defences of insects are inherently tied to the insect's ability to recognize the pathogen on several levels. First, pathogens may be entirely avoided if the insect can use its senses to proactively detect and avoid potential disease agents. Once the pathogen invades the insect, the ability to distinguish pathogens from "self" is critical to alerting the innate immune system of the host. Next, there are several physical barriers that deter entomopathogens from invading the host. The cuticle, peri-trophic matrix, and basal lamina are particularly effective at filtering the microbiota that infiltrates the haemocoel.

Cellular membranes are another barrier that limits intracellular pathogens from establishing infections. Entomopathogens must have adaptations that allow them to overcome these barriers if they are to survive in the host. Finally, the innate immune system possesses several weapons, including AMPs and the chemicals associated with the phenoloxidase cascade, haemocytes capable of phagocytosis and encapsulation, and xenophagy and RNA-based defences designed to kill intracellular pathogens. The pathogen must be able to hide from or disable the defences, or offer counter-defences in order to persist and reproduce in the host.

5.1. Introduction

Just like humans and other vertebrates, insects are susceptible to many disease-causing organisms known as pathogens. Thousands of species of bacteria, fungi, viruses, protozoa and nematodes can sicken or kill insects. Even if the insects survive, the pathogens' "sub-lethal" effects can keep their victims from feeding or reproducing. (Lundgren and Jurat-Fuentes, 2012).

Microorganisms that are capable of causing disease are called pathogens. The terms "infection" and "disease" are not synonymous. An infection results when a pathogen invades and begins growing within a host. Disease results only if and when, as a consequence of the invasion and growth of a pathogen, tissue function is impaired. Host bodies have defence mechanisms to prevent infection and, should those mechanisms fail, to prevent disease after infection occurs. Some infectious agents are easily transmitted (that is, they are very contagious), but they are not very likely to cause disease (that is, they are not very virulent).

Insects protect themselves against these detrimental microorganisms in two ways. Firstly, the body-enveloping chitin cuticle, in conjunction with the tracheal system and the gut lining, serve as major physical barriers to entry. Secondly, haemolymph itself plays a key role in repairing external wounds as well as destroying foreign organisms within the body cavity.

The effectiveness of these defences and how they evolve and spread throughout a population form the basis of resistance management plans that seek to preserve entomopathogen-based control methods for as long as possible. Insect defences, and how we can predict and delay their effectiveness, are the focus of this chapter.

5.2. Behavioural and Physical Barriers to Infection

Before encountering the immune responses of the host, the pathogen must breach several levels of defences that reduce both the encounters between the host and pathogen and the likelihood that the pathogen will reach the haemocoel. Reducing pathogen encounters and establishment involves insect behaviour in the case of avoidance and antiseptic practices, and it comprises the use of physical barriers that restrict the entrance of the pathogen. It is important to note that these physical barriers may be fortified by antimicrobial compounds, regulated by either the innate immune response or other sources.

5.2.1. Behavioural Avoidance

The most effective defence against disease is avoidance of contact with the disease-causing agent, which has been observed for some insect species (Alma *et al.*, 2010; Ormond *et al.*, 2011). Some examples are:

Case study : The omnivorous pirate bug, *Anthocoris nemorum*, avoids foraging on nettle leaves containing spores of *Beauveria bassiana*. Even more striking is that this bug avoids ovi-positing on tissue when it contains the fungus, thereby reducing the risk to its offspring (Meyling and Pell, 2006).

Case study: The gypsy moth, *Lymantria dispar*, detects cadavers and foliage that contain nucleopolyhedrovirus (NPV), and avoids these potential sources of occlusion bodies (Parker *et al.*, 2010).

In addition to being able to detect pathogenic organisms, some insects can detect and avoid some of the toxins produced by entomo-pathogens. For example:

Case study: Larvae of the beet armyworm, *Spodoptera exigua*, avoid consuming diet that contains Cry1C toxin, one of the lethal agents expressed by *Bacillus thuringiensis* (Berdegue´ *et al.*, 1996).

Most insects must come in contact with the pathogen or a cadaver containing the pathogen in order to recognize it as something to be avoided (Meyling and Pell, 2006; Thompson *et al.*, 2007; Alma *et al.*, 2010), but this is not always the case. For example:

Case study: The termite *Macrotermes michaelseni* not only can detect and avoid sources of *B. bassiana* and *Metarhizium anisopliae*, but also assesses their virulence from a distance and is more strongly repelled by the more virulent strains (Mburu *et al.*, 2009).

5.2.2. Antiseptic Behaviour

Simple grooming behaviour is effective in removing many pathogens from the exterior surface of the insect's body before they can cause disease. Although nearly all insects groom themselves in one way or another, many have specialized combs and brushes for just such behaviour. Social insects like bees, ants and locust, have limited genetic variability within a colony; they live in close quarters with their nest mates, and frequently store food in their nest. These characteristics make social insects particularly prone to diseases, but in reality these insects are fairly resistant to non-specialized pathogens. In addition to innate immunity, social insects also practice social immunity, or those behaviours that reduce the exposure of nest mates to potential pathogens (Cremer *et al.*, 2007; Stow and Beattie, 2008).

There is a wide range of examples of this social immunity in insects. Grooming in social insects is divided into Auto-grooming and Allo-grooming, depending on whether the insect is grooming itself or its nest mate (Evans and Spivak, 2010). Hygienic behaviour is another form of social immunity that is well studied in honey bees. For example:

Case study: In Hygienic behaviour, the workers bees detect larvae that are infected with the fungus *Ascosphaera apis* (chalkbrood disease of honey bees) or the bacterium *Paenibacillus larvae* (American foulbrood disease of honeybee) and they uncap the larval cell, remove the infected larva, and deposit it outside the nest (Evans and Spivak, 2010).

In another social insect example, ants practice necrophagy, which is similar to hygienic behaviour, except that the ants sometimes deposit their dead within isolated chambers of the nest (Evans and Spivak, 2010). Both hygienic behaviour and necrophagy are carried out by older members of the nest that typically do not care for brood, thereby limiting the exposure to their progeny (Evans and Spivak, 2010). If a nest mate becomes infected with a pathogen, for example,

Case study: When an ant colony of *Lasius neglectus* was infected with *M. anisopliae*, the workers instantly changed their behaviours to avoid contact with the brood and the uninfected ants were more likely to develop resistance to the pathogen after exposure to an infected nest mate (Ugelvig and Cremer, 2007).

5.3. Morphological Barriers to Infections

In addition to avoidance and behavioural removal of pathogens, insects possess a variety of morphological characteristics that present barriers to the establishment of pathogens. The precise barriers encountered depend on a pathogen's mode of transmission. The cuticle, tracheal system, and mid-gut are the major sites of invasion, and each of these fronts has its own set of physical defences.

5.3.1. Cuticle

The physical and inherent chemical characteristics of the cuticle have a great influence on which pathogens are able to establish on the external surface of a host. The cuticle covers all structures that come in contact with the external environment, including the exterior of the insect, the foregut, hindgut, and tracheal system (Chapman, 1998). The integument offers physical protection from predators, functions as an exoskeleton for muscle attachment, and reduces water loss, which allowed insects to permeate terrestrial ecosystems (Moussain, 2010). Another function of the integument is that it resists pathogen establishment.

Fungi are the main entomo-pathogens using the cuticular integument as a point of entry and start of infection. The entire external covering of insects is coated in a waxy layer, and the fatty acids that comprise this layer dictate which fungal entomo-pathogens are able to establish. Some fatty acids enhance the ability of certain fungi to adhere and germinate on the cuticle, while others restrict these processes (Bogus *et al.*, 2010). So far, it appears that fungi respond differently to the various fatty acids encountered on insect cuticles, so the insect needs to possess a wide variety of these fatty acids to have the plasticity necessary to repel the diversity of fungi they encounter. The exact mechanisms for how these fatty acids affect fungal adhesion and germination remain to be fully resolved.

Other factors, namely gland secretions and the microbial community, add antibiotic properties to the cuticle, making it an even more formidable barrier to infection. Some insects coat themselves in gland secretions that have antimicrobial properties. For instance,

> **Case study:** Ants cover themselves in an antibiotic milieu produced by thoracic meta-pleural glands, which are unique to Formicidae (Poulsen *et al.*, 2003; Stow and Beattie, 2008).

5.3.2. Digestive and Tracheal Systems

Many pathogens cannot penetrate the external defences of the insect host, and circumvent this barrier by attempting to enter the haemocoel through the mouth or tracheal system. These pathogens (many viruses, bacteria, microsporidia, and protists are noteworthy examples) are confronted by a series of formidable defences that necessitate specialized adaptations by the pathogen before successful infection is achieved. Much of the battle is waged in or near the mid-gut, where the cuticle is absent.

Microbes that enter the midgut must be adapted to survive and infect under the environmental conditions present in the particular insect, and these conditions vary widely among hosts. The pH, chemical milieu, and resident microbial community all play a role in which microbes are able to penetrate into the haemocoel. For example,

> **Case study:** *B. thuringiensis* pro-toxins are only activated under specific pH and with specific proteases, which for some extent narrows the host ranges of the various strains and toxins of this pathogen (Haider *et al.*, 1986).

Also, the salivary glands of some insects may produce antimicrobial compounds that permeate throughout the gut (Chouvenc *et al.*, 2010). Finally, the resident microbial community within insect guts resists some invaders in trying to reach a homeostatic condition, and even ordinarily pathogenic microbes can subsist as benign or even beneficial components of a healthy gut microbial community (Dillon and Dillon, 2004). The conditions fluctuate widely depending on the physiological status of the host and the environment in which this host lives.

If a microbe finds the environment of the gut suitable, the next barrier to infection is the peritrophic matrix. The peritrophic matrix is a semi-permeable, non-cellular envelope that encases the lumen of the mid-gut, and offers a barrier against physical abrasion by food and microbial invaders, such as bacteria and viruses. This membrane is primarily composed of microfibrils, but also proteo-glycans, proteins, and glycol-proteins, and acts as a filter for microorganisms seeking access to the epithelial cells of the mid-gut lining (Lehane, 1997).

5.4. Physiological Response to Infections

If a pathogen successfully navigates the external defences meant to keep it from becoming established, it could be faced with the defensive responses found within the insect hemocoel. Traditionally, the physiological responses of insects have been categorized as humoral or cellular in nature, although in reality, the two systems are intertwined in many aspects.

Innate physiological immunity follows a series of pathways to kill and eliminate pathogens from the insect's hemocoel. The first step in the process is that the insect must recognize potentially dangerous pathogens and alert the innate immunity systems. Once this recognition is achieved, one or more of several signalling pathways are initiated. The humoral response involves the production of antibiotics that kill the pathogen. Simultaneously, a cellular response could be initiated, whereby *hemocytes* aggregate around the pathogen, immobilizing and killing it. Finally, these two defense systems often work together to produce *melanin*, a third deadly weapon targeted against the extracellular pathogens in the haemocoel. The brief account of these response systems are mentioned hereunder.

5.4.1. Humoral Response System: During humoral responses to pathogens the main weapon that insects use is soluble antimicrobial peptides.

Antimicrobial Peptides

A number of antimicrobial peptides (AMPs) have been identified till date, and these peptides can be very specific for specific pathogens, or classes, thereof (Hoffmann, 2003; Govind, 2008; Kanost and Nardi, 2010). Lysozyme was the first AMP identified, and it was isolated in *Galleria mellonella* (Vilcinskas, 2010).

Many of the AMPs fall within one of three categories, based on their structure (Bulet *et al.*, 1999).

1. The first group of these is peptides with intramolecular disulfide bonds that form hairpin β-sheets and α-helical-β-sheet mixed structures.
2. The second group of AMPs has peptides that form amphipathic-heliceps
3. The third group has a disproportionate amount of proline and/or glycine residues.

There are various classes of AMPs within these groups. For example, *Cecropins* and *Defensins* are two major classes of AMPs specific for Gram-positive or Gram-negative bacteria, respectively (Hoffmann, 1995). *Proline-* and *glycine*-rich polypeptides are another class of AMP, and these are specific for Gram-negative bacteria. Acute phase *glycoproteins dTEPs* are another noteworthy groups of AMP that disrupt bacterial cell walls (Govind, 2008; Strand, 2008). In short, new antimicrobial peptides are frequently revealed, and our understanding of this rapidly expanding source of innate immunity would benefit from some synthesis in the future.

Regulation of Humoral Defences

Upon encountering a foreign target, signal transduction pathways that resist different pathogen classes is triggered (Govind, 2008). Although there are a number of pathways but usually one of these two pathways is triggered :

1. Toll pathway
2. IMD pathway

For example: fungi and Gram-positive bacteria, both activate the Toll pathway (Hoffmann, 2003). Both the Toll and IMD pathways are triggered upon exposure to certain viruses (Sabin *et al.,* 2010). The immunity provided by AMPs generally lasts for two to three days (Kanost and Nardi, 2010). Although it is known that they block the growth of microbes in the haemolymph (Hoffmann,

2003), the mechanisms by which the AMPs kill foreign targets remain poorly understood.

5.4.2. Cellular Response System

Cellular defence system of insects is inherently tied to humoral based immunity. Cellular defence system involves various types of hemocytes.

Haemocytes: Haemocytes kill and remove pathogenic targets through phagocytosis of small or unicellular targets, or through nodulation/encapsulation of larger elements, groups of pathogens, or multicellular organisms. All types of haemocytes are derived from prohaemocytes, which are a type of stem cell (Kanost and Nardi, 2010). Haemocytes are produced during two life stages of the insect, during embryogenesis and during late larval development (Strand, 2008).

Phagocytosis: The word "phagocytosis" has a Greek origin, which means "cell eating". In phagocytosis, granulocytes adhere to a foreign molecule or cell and envelop the invader, thereby killing it. During phagocytosis, the pathogen is encased intra-cellularly in a phagosome. This process is constrained by the size of the hemocyte and pathogen involved, and the cellular immune system must overcome larger invaders or groups of invaders using other means.

Nodulation and Encapsulation: Insects rely on nodulation and encapsulation to ensnare larger invaders that cannot be killed with a single granulocyte (Vilcinskas, 2010). Different insects produce different patterns in how the haemocytes aggregate to infections. Sometimes, granulocytes form the initial sheath around the target, and plasmatocytes and granulocytes build upon this. In other species, plasmatocytes and granulocytes arrange randomly, and granulocytes are not ubiquitously necessary for encapsulation to occur. Encapsulation is almost always reinforced by melanisation, a process discussed further below. Encapsulation almost always kills the target, either through asphyxiation or through the production of toxic chemicals during the melanisation process.

Melanization: It is a series of chemical reactions that results in the production of darkened pigment around a wound or pathogen, and there are three general steps in this chemical process (Carton *et al.*, 2008).

1. The first step is the activation of the phenoloxidase cascade, which occurs within minutes of infection (Royet, 2004b).
2. The next step is the proliferation and physical alterations of haemocytes at the site of the wound or invader.
3. The final step is the production of toxic chemicals that kills the offending agent.

The Melanization process produces several chemicals that are known to be toxic to microbes. Specifically, ROIs (O_2,H_2O_2), RNIs (NO), Quinones, hydroquinones, and most importantly, $_L$-DOPA are produced (Christensen *et al.*, 2005; Carton *et al.*, 2008).

5.5. Future Research Directions

There is no shortage of questions that can be confronted in future research efforts. Immuno challenged insects upregulate hundreds of genes, and in most cases we know almost nothing regarding their function (Baton *et al.*, 2008). Thus, although our understanding of host defences as a branch of insect pathology is growing rapidly, our concept of immune responses may still be somewhat limited. One topic that has received a fair amount of research, but little synthesis, is that insects and pathogens live in communities, and we know very little about how community interactions affect host resistance and defensive capabilities against pathogens. A future frontier is understanding the interaction between the biological communities and their effect on insect resistance. There is a growing understanding of the physiology and genetics that control insect pathogen interactions, and this understanding will provide the basis for asking more complex questions about how other organisms with which insects live affect the dynamics of insect pathology.

References

Alma, C.R., Gillespie, D.R., Roitberg, B.D. and Goettel, M.S. (2010). Threat of infection and threat-avoidance behaviour in the predator *Dicyphus hesperus* feeding on whitefly nymphs infected with an entomopathogen. *J. Insect Behav*. 23: 90–99

Altieri, M.A., Nicholls, C.I. and Fritz, M.A. 2005. *Manage insects on your farm: a guide to ecological strategies* (No. 632.96 A468m). Sustainable Agriculture Research and Education (SARE). Handbook series book 7, pp. 135. ISBN 1-888626-10-0.

Baton, L.A., Garver, L., Xi, Z. and Dimopoulos, G. (2008). Functional genomics studies on the innate immunity of disease vectors. *Insect Sci.,* 15: 15–27.

Berdegue, M., Trumble, J.T. and Moar, W.J. (1996). Effect of CryIC toxin from *Bacillus thuringiensis* on larval feeding behaviour of *Spodoptera exigua. Entomol. Exp. Appl.*, 80: 389–401.

Bogus, M.I., Czygier, M., Go1ebiowski, M., Kedra, E., Kucinska, J., Mazgajska, J., Samborski, J., Wieloch, W. and Wloka, E. (2010). Effects of insect cuticular fatty acids on *in vitro* growth and pathogenicity of the entomopathogenic fungus Conidiobolus coronatus. *Exp. Parasitol.*, 125: 400–408.

Bulet, P., Hetru, C., Dimarcq, J.L. and Hoffmann, D. (1999). Antimicrobial peptides in insects; structure and function. *Dev. Comp. Immunol.*, 23: 329–344.

Carton, Y., Poirie, M. and Nappi, A.J. (2008). Insect immune resistance to parasitoids. *Insect Sci.,* 15: 67–87.

Chapman, R.F. (1998). The Insects: Structure and Function (4th *Ed.*). Cambridge: Cambridge University Press.

Chouvenc, T., Su, N.Y. and Robert, A. (2010). Inhibition of the fungal pathogen *Metarhizium anisopliae* in the alimentary tracts of five termite (Isoptera) species. *Fla. Entomol.,* 93: 467–469.

Christensen, B.M., Li, J., Chen, C.C. and Nappi, A.J. (2005). Melanization immune responses in mosquito vectors. *Trends Parasitol.,* 21: 192–199.

Cremer, S., Armitage, S.A. and Schmid-Hempel, P. (2007). Social immunity. *Curr. Biol.*, 17: R693–R702.

Dillon, R.J. and Dillon, V.M. (2004). The gut bacteria of insects: non-pathogenic interactions. *Annu. Rev. Entomol.*, 49: 71–92.

Evans, J.D. and Spivak, M. (2010). Socialized medicine: individual and communal disease barriers in honey bees. *J. Invertebr. Pathol.*, 103: S62–S72.

Gillott, C. (1995). *Entomology* (2 Edn.). New York: Plenum Press. ISBN 0-306-44967-6.

Govind, S. (2008). Innate immunity in Drosophila: pathogens and pathways. *Insect Sci.*, 15: 29–43.

Haider, M.Z., Knowles, B.H. and Ellar, D.J. (1986). Specificity of *Bacillus thuringiensis* var. *colmeri* insecticidal d-endotoxin is determined by differential proteolytic processing of the protoxin by larval gut proteases. *Eur. J. Biochem.*, 156: 531–540.

Hoffmann, J.A. (2003). The immune response of Drosophila. Nature, 426: 33–38.

https://en.wikipedia.org/wiki/Defense_in_insects

Kanost, M.R. and Nardi, J.B. (2010). Innate immune responses of *Manduca sexta*. In M.R. Goldsmith and F. Marec (Eds.), *Molecular Biology and Genetics of the Lepidoptera* (pp. 271–291). Boca Raton: CRC Press.

Lehane, M.J. (1997). Peritrophic matrix structure and function. *Annu. Rev. Entomol.*, 42, 525–550.

Lundgren, J.G. and Jurat-Fuentes, J.L. (2012). Physiology and ecology of host defense against microbial invaders. In *Insect Pathology (Second Edition)* (pp. 461–480).

Mburu, D.M., Ochola, L., Maniania, N.K., Njagi, P.G.N., Gitonga, L.M., Ndung'u, M.W., Wanjoya, A.K. and Hassanali, A. (2009). Relationship between virulence and repellency of entomo- pathogenic isolates of *Metarhizium anisopliae* and *Beauveria bassi- ana* to the termite *Macrotermes michaelseni*. *J. Insect Physiol.*, 55: 774e780.

Meyling, N.V. and Pell, J.K. (2006). Detection and avoidance of an entomopathogenic fungus by a generalist insect predator. *Ecol. Entomol.*, 31: 162–171.

Moussain, B. (2010). Recent advances in understanding mechanisms of insect cuticle differentiation. *Insect Biochem. Mol. Biol.*, 40: 363e375.

Ormond, E.L., Thomas, A.P.M., Pell, J.K., Freeman, S.N. and Roy, H.E. (2011). Avoidance of a generalist entomopathogenic fungus by the ladybird, *Coccinella septempunctata*. *FEMS Microbiol. Ecol.,* 77: 229–237.

Parker, B.J., Elderd, B.D. and Dwyer, G. (2010). Host behaviour and exposure risk in an insect pathogen interaction. *J. Anim. Ecol.*, 79: 863–870.

Poulsen, M., Bot, A.N.M. and Boomsma, J.J. (2003). The effect of met pleural gland secretion on the growth of a mutualistic bacterium on the cuticle of leaf-cutting ants. *Naturwissenschaften,* 90: 406–409.

Royet, J. (2004b). Drosophila melanogaster innate immunity: an emerging role for peptidoglycan recognition proteins in bacteria detection. *Cell. Mol. Life Sci.*, 61: 537–546.

Sabin, L.R., Hanna, S.L. and Cherry, S. (2010). *Innate antiviral immunity in Drosophila.* Curr. Opin. Immunol., 22: 4–0.

Stow, A. and Beattie, A.J. (2008). Chemical and genetic defences against disease in insect societies. *Brain Behav. Immun.*, 22: 1009–1013.

Strand, M.R. (2008). The insect cellular immune response. *Insect Sci.*, 15: 1–14.

Thompson, S.R., Brandenburg, R.L. and Roberson, G.T. (2007). Entomopathogenic fungi detection and avoidance by mole crickets (Orthoptera: Gryllotalpidae). *Environ. Entomol.* 36: 165–172.

Ugelvig, L.V. and Cremer, S. (2007). Social prophylaxis: group interaction promotes collective immunity in ant colonies. *Curr. Biol.*, 17: 1967–1971.

Vilcinskas, A. (2010). Lepidopterans as model mini-hosts for human pathogens and as a resource for peptide antibiotics. In: M.R. Goldsmith & F. Marec (eds.), *Molecular Biology and Genetics of the Lepidoptera* (pp. 293–306). Boca Raton: CRC Press.

Question time

1. Discuss in brief the Behavioural and Physical Barriers to Infection in insect host.
2. Quote few examples of Behavioural Avoidance in insects.
3. Discuss in brief the Antiseptic Behaviour or Hygienic behaviour in the honey bees.
4. Describe in details the Cellular Response System of an insect against pathogen.

Chapter 6

MASS PRODUCTION OF MICROBIAL PESTICIDES

In nature every ecosystem exists in a balance. Growth and multiplication of each organism depends on the food-chain, its predators, parasites, etc. In biological control system, these inter-relations are exploited. The natural enemy of a pest, disease or weed is selected; its biology is studied for mass multiplication and utilizes the same to check the target pest. They are also specific in their action and perish once their feed (*i.e.* the pest) is exhausted. Thus they are based on natural principles; do not leave any residue, safe and economical. Among the alternatives, biological control of pests is one of the important means for checking pest problems in almost all agro-ecological situations.

Bio-pesticides are living organisms which can intervene the life cycle of insect pests in such a way that the crop damage is minimized. The agents employed as bio-pesticides, include parasites, predators and disease causing pathogens (*viz.* fungi, bacteria and viruses) which are the natural enemies of pests. Further, they complement and supplement other methods of pest control. Utilisation of naturally occurring parasites, predators and pathogens for pest control is a classical biological control. On the other hand, these bio-agents can be conserved, preserved and multiplied under laboratory condition for field release. Once these bio-agents are introduced in the field to build their population considerably, they are capable of bringing down the targeted pest population below economic threshold level (ETL). However, the crux lies in their mass production and application at the appropriate time.

Scope for Commercial Production of Bio-pesticides

Though there are about 140 bio-pesticide production units existing in the country as on today, they are able to meet the demand of only less than 1 per cent of cropped area. There exists a wide gap, which can only be bridged by setting up of more and more units for production of bio-pesticides. This requires large scale investment and private participation. In India the technology used for mass production is indigenous and the scientific aspects of production were standardised by ICAR Research Institutes and State Agricultural Universities. Machineries and laboratory equipments are available from various manufacturers and are of BIS standards.

Objectives of Bio-pesticide Mass Production Project in India

- The primary objective of bio-pesticide projects is to establish the bankability of mass multiplication of various bio-agents discussed in the models.
- To serve as guidelines for extending financial assistance to entrepreneurs who may be interested in setting up bio-pesticide units.
- To promote setting up of more bio-control production units.
- To disseminate the technology widely.

Basic Requirements for Establishment of Bio-pesticide units

Based on the field visits to bio-control production units and in line with the technology and objective of bio-pesticides production, various facilities required for the successful implementation of such projects are indicated below:

1. Land

Land is required for construction of culture and rearing rooms, processing room, laboratory, office etc. For this purpose an estimated built up area of about 2400 sq. feet is required for production of NPV, fungal or bacterial insect pathogens.

2. Building and Civil Works

Bio-pesticides production involves rearing of insects, separate room for diet preparation, egg production, host culture etc. Hence, the basic infrastructure to be created includes civil structures built in such a way as to provide environmental conditions suitable for rearing of insects. The production unit has to be located away from industrial unit to avoid pollution problems. Other utilities required are power, water and vehicle.

3. Plant and Machinery

There is no requirement of heavy plant and machinery. Racks, trays and other facilities are required for rearing insects. Apart from this centrifuge, mixers and some fabricated equipments for insect collection and rearing are required.

4. Raw Material

For rearing of insects, special diet is required which comprises of pulses, vitamins, antibiotics etc. All these materials are available locally.

5. Water

The water requirement is mainly for feed preparation, washing, cleaning, drinking etc. Water quality should be tested to establish the suitability.

6. Power

Power supply is essential for bio-pesticide units. Electricity charges under recurring cost are considered in the models.

7. Location of Bio-pesticide Units

In order to achieve optimum results, care needs to be taken to set up bio-pesticide facilities in areas which have appropriate climatic conditions. Temperature control is less costly in locations where there are no extreme conditions. Besides the climatic conditions, the proximity of the location to the market is also important. However, care must be taken that the production facilities are set up at least a quarter of a mile away from farming areas, so as to prevent the contamination of production facilities by insecticides from the farming areas. Also, as air pollution can damage bio-pesticides, the production should be located away from industrial and urban areas.

8. Manpower

Production of bio-pesticides required skilled manpower. There is need for a number of labourers at each stage of production. The project is labour intensive. The manpower requirement is as under:

Designation	Manpower
Executive Officer	1
Chief Biologist/Micro-Biologist	1
Sales Officer	1
Accountant and clerical Assistant	1
Drivers	1–2
Floor Supervisor/Factory Manager	1
Technical Staff (boiler operation, mechanical maintenance, packing machine operations, electrical maintenance)	12
Skilled labourers	2–3
Semi-skilled, depending upon the volume of production	3–5
Unskilled	3

9. Bio-pesticides Registration

At present, *Bacillus thuringensis*, neem based formulations, microbial pesticides like fungi, NPV etc., are included in the schedule of Insecticides Act, 1968. This ensures the quality of bio-pesticides at farmers' level. The standard parameters, protocols for data generation, guidelines for registration are prepared and circulated to prospective entrepreneurs. Now as such, any person dealing with bio-pesticides without registration is illegal.

10. Technical Consultancy for Setting up of Mass Production of Bio-agents

Setting up of unit for mass production of bio-agents especially those which are based on fungi, bacteria and virus is highly technical in nature. The skill required to handle the mass production process is also higher. For scientific and successful setting up of a unit, the entrepreneurs can take consultancy services from the following agencies:

- Project Directorate, Biological Control, ICAR, Bangalore.
- Indian Institute of Horticulture Research, Hesaragatta, Bangalore.
- Central Integrated Pest Management Centre (CIPMC), White field, Bangalore.
- Central Institute for Cotton Research (CICR), Nagpur.

11. Capital Investment Subsidy Scheme (CISS) for Bio-fertilizer/bio-pesticide Production Units

25 per cent of total financial outlay subject to the maximum of INR 40 lakh per unit, whichever is less. NABARD releases subsidy to the units financed by:

- Commercial Banks,
- Regional Rural Banks (RRBs),
- State Cooperative Banks (SCBs),
- State Cooperative Agricultural and Rural Development Banks.

Production Technologies of Microbial Pesticides

Early in the development of any microbial pesticides product there is a requirement for relatively small quantities of the microbes for both laboratory and field testing. The methods of production vary according to the experimental design and the stage of development within the project. For bioassays and general isolate screening, maintenance and production on agar slants in bottles is usually sufficient. As the project develops and one or more isolates have been selected for further development, greater amounts of material are required and it becomes necessary to demonstrate the feasibility of large-scale production. This requires the development of a production system which not only produces microbial pesticides in sufficient quantities for experimental use, but which also are of consistent quality and compatible with the selected formulation and application equipment.

A feature of the majority of research and development in mass production programmes is that large amounts of capital are not available for the design and construction of an industrial scale production system. The development and optimization of a simple production procedure will not only be invaluable for the supply of reliable material for laboratory and field efficacy trials, but will help to ensure the compatibility of the product with the formulation and application system. A tremendous amount can be learnt about the microbial isolate being developed, permitting its production characteristics, stability and long-term storage to be established well in advance of eventual commercial development and production.

This chapter describes the basics of developmental processes involved in mass production of various insect pathogens under following heads.

1. Fungal pathogens
2. Viral pathogens
3. Mass Production of other Bio-control agents

1. Mass production of fungal pathogens

Fungi represent a diverse group of insect pathogens. The insects attacked by the fungus die shortly after the fungus begins to develop in the haemocoel.

- Mass production of green muscardine fungus (*Metarhizium anisopliae)*
- Mass production of white muscardine fungus (*Beauvaria bassiana)*
- Mass production of white halo fungus (*Verticillium lecanii)*

A. Mass production of green muscardine fungus - *Metarhizium anisopliae*

Case study: The Rhinoceros beetle, *Oryctes rhinoceros* is one of the serious and important pests of coconut and has a wide distribution and persistent occurrence in all the coconut growing areas in the country. The adult beetle cause severe damage to the growing palms by feeding on the tender fronds and crowns and resulting in stunting of the trees. The damage to the spathe results in the loss of nuts. Young seedlings are sometimes killed outright.

Since the insect breeds in the farmyard manure and fallen coconut trees, the control measures have to be directed at the breeding sites as well as on the trees. The chemical control measures adopted against this pest are always costly, tedious and have to be repeated.

Hence an easy method utilizing the safe and specific fungus, *Metarhizium anisopliae* for the management of this pest is found suitable.

Green muscardine fungus

- The colony of *M. anisopliae* appears white when young, but as the conidia mature, the colour turns to dark green.
- The conidiophores are branched, and the initial conidium is produced at the distal end of the conidiophores.
- A chain of conidia is formed on each conidiophore with the youngest conidium being adjacent to the conidiophore.

- The mass of spore chains becomes so dense and coheres with each other to produce prismatic masses of columns of spore chains.

In Carrot broth

- Carrot cut into small pieces (40 g) is washed in potable water and transferred to conical flask (250 ml) and 15 ml of distilled water is added.
- The conical flasks are plugged with cotton and autoclaved for 20 min at 15 psi.
- The flasks are allowed to cool and taken to laminar flow chamber for inoculation.
- From a clean uncontaminated mother culture in slant, loopful quantities of *M. anisopliae* spores are transferred aseptically.
- The flasks are incubated at room temperature.
- The spores can be harvested in a fortnight.

B. Mass production of White muscardine fungus - *Beauvaria bassiana*

Introduction

- The fungus spores and mycelia are milky white and found sprouting on the body of Lepidopterans insects like *Helicoverpa armigera*, *Spodoptera litura* and *Anadevidia peponis*.

Production procedure

- Carrot cut into small pieces (40 g) is washed in potable water and transferred to conical flask (250 ml) and 15 ml of distilled water is added.
- The conical flasks are plugged with cotton and autoclaved for 20 min at 15 psi. The flasks are allowed to cool and taken to laminar flow chamber for inoculation.
- From a clean uncontaminated mother culture in slant loopful quantities of *B. bassiana* spores are transferred aseptically.
- The flasks are incubated at room temperature.
- The spores are obtained in a fortnight.

C. Mass production of White halo fungus - *Verticillium lecanii*

Introduction

- *Vertilcillium lecani* commonly called as white halo fungus is found sprouting on the body of coffee green scale *Coccus viridis*.

- It has been extensively studied against the pest in Lower Pulney Hills in Tamil Nadu.
- The fungus is known to cause epizootic when the environmental conditions are favourable.

Production procedure

- The fungus is multiplied on cheap media for large scale production.
- Sorghum grains devoid of pesticide residues (40 g) is washed in potable water and transferred to conical flask (250 ml) and 15 ml of distilled water is added.
- The conical flasks are plugged with cotton and autoclaved for 20 min at 15 psi.
- The flasks are allowed to cool and taken to laminar flow chamber for inoculation.
- From a clean uncontaminated mother culture in slant loopful quantities of *V. lecani* spores are transferred aseptically.
- The flasks are incubated at room temperature.
- The spores are obtained in a fortnight.

2. Mass production of viral pathogens

Mass production technique of some important viruses.

- Mass production of the Granulosis Virus of *Chilo infuscatellus.*
- Mass production of the nuclear polyhedrosis virus *Spodoptera litura*
- Mass production of the nuclear polyhedrosis virus of *Helicoverpa armigera.*

A. Mass production of the Granulosis Virus of *Chilo infuscatellus.*

Introduction

- *Chilo infuscatellus* is a devastating pest of sugarcane. Management of the pest with insecticides has become extremely difficult and the search for alternatives has yielded a Granulosis virus.
- The virus is effective when applied at 1.5×10^{13} OB/ha repeatedly three or four times on early cane under field condition.

Production procedure

- The Granulosis virus is produced *in vivo*. The method employed in production is by oral feeding treatment.

- Fresh bits of sugarcane shoots from field are used.
- The cut splits are washed and placed on moist filter paper in 6.6×7.4 cm containers @ 3 bits per container.
- A dose of 1×10^{8}OB/ml in Teepol (0.01%) is prepared and larvae in 4th instar collected from field are treated by dipping the head into the suspension containing the virus in a watch glass.
- The larvae @ 1–2/shoot bit are transferred to the container and the jar is closed. The feed is replaced with fresh ones every alternate day.
- The larvae begin to die after 6 days and are collected in sterile containers and stored in a freezer.
- The processing methods are similar to that of the nuclear polyhedrosis virus.
- In differential centrifugation, the cadavers are centrifuged at 700 rpm initially and at 8000 rpm for 30 min finally to collect the pellet.

B. Mass production of the nuclear polyhedrosis virus *Spodoptera litura*

Introduction

- The virus is effective against the pest on several crops.
- The virus is generally propagated in early fifth instar by surface contamination of semi-synthetic diet.

Production procedure

- The mass production of the NPV is carried out in early fifth instar stage of *S. litura*, which yields maximum amount of the NPV. Therefore, in the host culture laboratory a continuous culture of the insects is maintained with proper handling procedures.
- The larvae are grown in diet held in 5 ml glass vials. When the larvae reach the appropriate stage they are transferred to the virus production facility.
- The NPV is multiplied by feeding the semi-synthetic diet coated with a clean inoculum of the NPV that has previously been standardized.
- This is accomplished by placing aliquots of 10 ml of the viral suspension of concentration 1×10^{8} Polyhedral Occlusion Bodies (POB) in the centre over the diet surface, either in glass vials and spreading the suspension uniformly all over the surface with a polished glass rod.
- Larvae are released singly after 15 min. into each glass vial/cell and incubated at 25°C for 10 days.

- The larvae begin to die from 5th day onward. The cadavers are collected individually and transferred to 500 ml plastic containers and frozen immediately until processing.
- The processing method is similar to that of *H. armigera*.

C. Mass production of the nuclear polyhedrosis virus of *Helicoverpa armigera*

Introduction

In India, *Helicoverpa armigera* is of major importance damaging a wide variety of food, fibre, oilseed, fodder and horticultural crops. The nuclear polyhedrosis virus of *H. armigera* (HaNPV) is currently used for the management of *H. armigera* on chickpea, cotton, pigeon pea, tomato and sunflower. Mass production of Nuclear Polyhedrosis Virus (NPV) on commercial scale is restricted to *in vivo* procedures in host larvae which are obtained by

- Field collection from cotton, pigeon pea and chickpea – *H. armigera*
- Mass culturing in the laboratory in semi-synthetic diet – *H. armigera*

Some small scale producers use field collected larvae for mass production of NPV in spite of the following constraints.

- Collection of a large number of larvae in optimum stage (late IV / early V instars) is time-consuming and can be expensive in terms of labour and transportation costs.
- Wild populations of insects may carry disease causing organisms like microsporidians, cytoplasmic polyhedrosis virus, stunt virus and fungal pathogens which will affect both virus production and quality.
- Introduction of wild strains of NPV resulting in quality control problems.
- Transportation of a large number of larvae with cannibalistic behaviour will be a difficult task.
- Parasitized larvae collected from the field will die prematurely yielding little virus.

Rearing of larvae in the natural host plant will involve frequent change of food at least once a day during the incubation period of 5–9 days, increasing the handling time and hence the cost. In order to reduce the cost, field collected larvae are released into semi-synthetic diet treated with virus inoculum. Mass culturing of insects in semi-synthetic diet involves high level of expertise, hygiene and cleanliness.

Production procedure

- The NPV of *H. armigera* is propagated in early fifth instar larvae.

- The virus is multiplied in a facility away from the host culture laboratory.
- The dose of the inoculum used is 5×10^5 polyhedral occlusion bodies (POB) in 10 ml suspension.
- The virus is applied on to the semi-synthetic diet (lacking formaldehyde) dispensed previously in 5 ml glass vials.
- A blunt end polished glass rod (6 mm) is used to distribute the suspension containing the virus uniformly over the diet surface.
- Early fifth instar stage of larvae are released singly into the glass vials after inoculation and plugged with cotton and incubated at a constant temperature of 25°C in a laboratory incubator.
- When the larvae exhausted the feed, fresh untreated diet is provided.
- The larvae are observed for the development of virosis and the cadavers collected carefully from individual bottles starting from fifth day.
- Approximately, 200 cadavers are collected per sterile cheese cup (300 ml) and the contents are frozen immediately.
- Depending upon need, cadavers are removed from the refrigerator and thawed very rapidly by agitation in water.

Processing of NPV

- The method of processing of NPV requires greater care to avoid losses during processing.
- The cadavers are brought to normal room temperature by repeatedly thawing the container with cadaver under running tap water.
- The cadavers are homogenized in sterile ice cold distilled water at the ratio 1: 2.5 (w/v) in a blender or precooled all glass pestle and mortar.
- The homogenate is filtered through double layered muslin and repeatedly washed with distilled water.
- The ratio of water to be used for this purpose is 1:7.5–12.5 (w/v) for the original weight of the cadaver processed.
- The left over mat on the muslin is discarded and the filtrate can be semi-purified by differential centrifugation.
- The filtrate is centrifuged for 30–60 sec. at 500 rpm to remove debris.
- The supernatant is next centrifuged for 20 min at 5,000 rpm.
- Then the pellet containing the polyhedral occlusion bodies (POB) is suspended in sterile distilled water and washed three times by centrifuging the pellet in distilled water at low rpm followed by centrifugation at high rpm.

- The pellet finally collected is suspended in distilled water and made up to a known volume, which is necessary to calculate the strength of the POB in the purified suspension.

3. Mass Production of other Bio-Control agents

- Mass production of *Trichoderma viride*
- Mass production of *Pseudomonas fluorescens*
- Mass multiplication of *Bacillus subtilis*
- Preparation of talc based products

A. Mass production of *Trichoderma viride*

Preparation of mother culture

Molasses yeast medium is prepared as detailed below

Molasses	: 30 g
Yeast	: 5 g
Distiller water:	1000 ml

The medium is prepared and dispensed into conical flasks and sterilized at 15 lb pressure for 15 minutes in an autoclave. After the medium is cooled it is inoculated with 10 days old fungal disc of *T. viride* and then incubated for 10 days for fungal growth. This serves as mother culture.

Mass multiplication

- Molasses yeast medium is prepared in fermenter and sterilized as described earlier.
- Then after the medium is cooled, the mother culture is added to the fermenter @ 1.5 lit / 50 lit of the medium and incubated at room temperature for 10 days.
- Then the incubated broth containing the fungal culture is used for commercial formulation preparation using talc powder.

B. Mass production of *Pseudomonas fluorescens*

Preparation of mother culture

Mother culture is prepared by using the king's B medium

Peptone	: 20.0 g
K_2HPO_4	: 1.5 g
$MgSO_4$	: 1.5 g
Glycerol	: 10 ml

Distilled water : 1000 ml

The above broth is dispersed into conical flasks and autoclaved at 15 lb pressure for 15 minutes and cooled, and inoculated with a loop of *P. fluorescens* and incubated for 2 days.

Mass multiplication

- The kings B medium is prepared and poured into the fermenter and sterilized at 15 lb pressure for 15 minutes.
- After the broth has cooled below the mother culture of *P. fluorescens* is added to the king's B medium in the fermenter at the rate of 3 lit for 40 lit of the broth.
- Then it is incubated in the fermenter for 2 days with frequent mixing of the broth by operating the stirrer.
- Then the broth containing the bacterial growth is collected in plastic buckets and used for mixing with talc powder for commercial formulation.

C. Mass multiplication of *Bacillus subtilis*

Preparation of mother culture

The nutrient broth medium is prepared as detailed below

Glucose	: 5.0 g
Peptone	: 5.0 g
Beef extract	: 3.0 g
Sodium chloride	: 3.0 g
Distilled water	: 1000 ml

The above medium is dispensed in conical flasks and autoclaved at 15 lb pressure for 15 min. A loop of *B. subtilis* is inoculated into the medium and incubated for 2 days. This serve as the mother culture.

Mass multiplication

- The nutrient broth is prepared in fermenter and sterilized at 15 lb pressure for 15 min.
- Then the mother culture is added @ 1 lit / 100 lit of the medium and incubated at room temperature for 2 days.
- The medium containing the bacterial growth of *B. subtilis* is used for mixing with talc powder.

D. Preparation of Talc based products, Air drying of formulation and Estimation of Moisture content

(A) *Trichoderma viride*

- The fungal biomass collected from fermenter is mixed with talc powder at 1:2 ratio.
- The mixture is air dried in shade and mixed with carboxy methyl cellulose (CMC) @ 5 g / kg the product.
- It is packed in polythene bags and should be used within 4 months.

Quality control parameters:

1. Fresh product should contain hot less than 28×10^6 CFU/g
2. After 4 months of storage at room temperature, the population should be 20×10^6 CFU/g.
3. Maximum storage period in talc is 4 months.
4. The talc size should be 500 microns
5. The product should be packed in polythene bags
6. Moisture content of the final product should not be more than 20%

(B) *Bacillus subtilis*

- The broth containing the bacteria is collected from fermenters and mixed with 250 kg of sterilized neat soil for 100 lit of broth.
- Then 37 kg if calcium carbonate is added thoroughly mixed, dried is shade and packed in polythene bags.
- This can be stored up to 6 months.

C. *Pseudomonas fluorescens*

- The broth containing the bacterial growth is collected from fermenter and added @ 400 ml/kg of talc powder.
- Then CMC is added @ 5 g/kg mixed well air dried to 20% moisture level and packed in polythene bags.

Quality control parameters

1. Fresh produce should contain 2.5×10^8 CFU/g.
2. After 3 months of storage at room temperature the population should be $8–9 \times 10^7$ CFU/g.
3. Storage period is 3–4 months.
4. Minimum population load should be 1.0×10^8 CFU/g.

5. Moisture content should not exceed 20% in the final product.
6. Population per ml of the broth should be 2×10^8 CFU/g.

References

http://agritech.tnau.ac.in/crop_protection/crop_prot_bio_mass_fungus1.html

http://agritech.tnau.ac.in/crop_protection/crop_prot_bio_mass_fungus2.html

http://agritech.tnau.ac.in/crop_protection/crop_prot_bio_mass_fungus3.html

http://agritech.tnau.ac.in/crop_protection/crop_prot_bio_mass_virus_1.html

http://agritech.tnau.ac.in/crop_protection/crop_prot_bio_mass_virus_2.html

http://agritech.tnau.ac.in/crop_protection/crop_prot_bio_mass_virus_3.html

http://agritech.tnau.ac.in/crop_protection/crop_prot_bio_mass%20production_trichoderma.html

http://agritech.tnau.ac.in/crop_protection/crop_prot_bio_mass%20production_pseudomonas.html

http://agritech.tnau.ac.in/crop_protection/crop_prot_bio_mass%20production_Basicillus.html

http://agritech.tnau.ac.in/crop_protection/crop_prot_bio_mass%20production_talc_based_products.html

http://vibrantgujarat.com/writereaddata/images/pdf/project-profiles/Manufacturing-of-Biofertilizers-and-Biopesticides.pdf

Question time

1. What are Bio-pesticides? Discuss Scope for commercial production of Bio-pesticides in India.
2. Describe in brief the objectives of bio-pesticide mass production Project in India.
3. Describe in brief the Mass production of green muscardine fungus.
4. Describe in brief the Mass production of *Trichoderma viride.*
5. Describe in brief the Mass production of *Pseudomonas fluorescens.*
6. Describe in brief the Mass multiplication of *Bacillus subtilis.*
7. Describe in brief the Preparation of talc based products of bio-pesticide.

True or false

1. The insects attacked by the fungus die shortly after the fungus begins to develop in the haemocoel – **True**
2. The colony of *M. anisopliae* appears dark green at young and mature stages – **False**
3. The NPV of *H. armigera* is propagated in early fifth instar larvae – **True**

4. The Granulosis virus is produced *in vitro* - **False**
5. The nuclear polyhedrosis virus of *H. armigera* (HaNPV) is currently used on chickpea, cotton, pigeon pea, tomato and sunflower - **True**

■■■■

Chapter 7

SAFETY AND REGISTRATION OF MICROBIAL PESTICIDES IN INDIA

Where scathe is expected to human and environmental health from the insertion of a microbial pest control product, regulatory procedures for pre-market assessment of safety are common in most countries.

Safety of entomopathogenic microorganisms used in microbial control is a major concern associated with the increased needs for bio-control agents desired in the exploitation of IPM programs for agricultural, veterinary, and medical insect pests (Burges, 1981; Burges, *et al.*, 1980). Particular attention is paid to the following aspects by registration of the bio-products based on microbes: (1) allergic properties, (2) risks of toxic metabolites, (3) genetic recombination and displacement of natural strains, and (4) effect on biodiversity, *i.e.,* on non-target organisms. These aspects have been covered in depth in chapter 8.

An effective regulatory framework provides protection from harm, while still facilitating the availability of useful products. It recognizes the needs of farmers, society and the commercial interests of the pest control product proponents *i.e.* manufacturing companies. The product manufacturing companies are recognized by being provided clear communication of the terms of the regulatory system and the stepwise process of product registration, and there by being charged reasonable costs for registration, so products move seamlessly from the application for registration through to the market. The risks assumed with the introduction of microbial pest control products are related to their toxicity, infectivity, pathogenicity to and displacement of non-target organisms, and the potential irreversibility of introduction into the environment. These are the types of risks which are assessed during a pre-market safety evaluation (Cook *et al.*, 1996; Jaronski *et al.*, 2003; Mensink and Scheepmaker, 2007; Organization for Economic Cooperation and Development, 2003).

Regulatory systems for pest control products are designed to achieve several purposes. They must protect human health and the environment from potential risks associated with the use of pest control products. They must also allow nations to meet their needs for food and fibre production by facilitating the availability of pest control products and ensuring that products have acceptable efficacy.

An effective regulatory system achieves these goals by putting in place the following elements:

- A system of data requirements to guide the assessment of human health and safety, value (including efficacy), and environmental safety;
- Clear and predictable procedures for assessing the risk and value of pest control products, with sufficient flexibility to allow expert opinion to contribute to the assessment process;
- Mechanisms which afford opportunities for public and industry input into the decision making process, including the right to appeal decisions;
- Policies which establish reasonable timelines for assessment of various classes of products, and an agency with a good track record with regard to these timelines;
- The flexibility to modify regulatory procedures in line with new scientific information;
- Regulatory fees which are affordable to registrants (Manufacturing companies);
- Enforcement of legislation and regulations related to product use, sale, distribution, and other regulatory requirements.

Indian scientific interest in the use of bio-pesticides for crop protection is long standing, with active research into entomopathogenic fungi being pursued since the 1960s. However, it was probably the impact of the Indian agricultural crisis of the late 1980s, due to widespread failure of chemical insecticides to control *Plutella xylostella* in Cole crops; *Helicoverpa armigera*, *Spodoptera litura* and other pests in cotton (Armes *et al.* 1992; Kranthi *et al.* 2002), that prompted efforts to develop systematic integrated pest management (IPM) and insecticide resistance management programmes, of which bio-pesticides are a component.

The commercial production of bio-pesticides began in the 1980s, but expansion became rapid in the late 1990s stimulated by national and state programmes for IPM promotion (Wahab 2004). Other bio-pesticides currently under development include *Hyblea puera* NPV for controlling teak defoliator (Biji *et al.*, 2006), *Amsacta albistriga* NPV for controlling this pest on groundnuts (Veenakumari *et al.*, 2007), *Nomurea rileyi, Pochomia chlamydosporia,* and entomopathogenic nematodes.

It has been estimated that there are at least 32 commercial companies active in bio-pesticide production, with an additional 32 IPM centres under the Ministry of Agriculture also producing selected bio-control agents (Singhal, 2004). The state departments of agriculture and horticulture in the states of Tamil Nadu, Kerala, Karnataka, Andhra Pradesh, Chhattisgarh and Gujarat have established biocontrol laboratories for producing selected microbial bio-control agents. A few state agricultural universities (*viz.* Indira Gandhi Agriculture University, Raipur Chhattisgarh) and Indian Council of Agricultural Research (ICAR) institutions also produce small quantities of microbial pesticides (Rabindra,

2005). In total, at least 410 bio-pesticide production units have been established in India, 130 in the private sector (Singhal, 2004).

The microbial pesticides used in India are listed in Table 6.1 (According to the recommendation of Government of India, Ministry of Agriculture – Major use of pesticide: Registered under the Insecticide Act, 1968. As on 31.08.2015). The products registered, including *Bacillus thuringiensis* (*Bt*), are sourced partly as imported products, but also include many that are locally produced. Indian bio pesticide production is currently dominated by antagonistic fungi and bacteria such as *Trichoderma* sp. and *Pseudomonas fluorescens, Beauveria bassiana, Metarhizium anisopliae, Verticillium lecanii* etc. (Table 6.1), but the production of nucleopolyhedrosis viruses (NPV), granuloviruses (GV), and entomo-pathogenic fungi are also established and expanding (Rabindra, 2005).

Table 6.1. MAJOR USES OF PESTICIDES (Registered under the Insecticides Act, 1968) UPTO- 31.08.2015

Sr no.	Microbial Insecticides	Texas ingre-dients	Dosage / ha	Target	
			Formulation (gm/ml)	Crops	Pest
1.	*Bacillus thuringiensis* var. *galleriae*	Bacterium	06–1.0 kg	Cabbage & Cauliflower	Diamond back moth *(Plutella xylostella)*
			1.0–1.5 kg	Tomato	Fruit borer *(H. armigera)*
			1.0–1.5 kg	Okra	Fruit borer (*Earias* sp.)
			1.5–2.0 kg	Chillies	Fruit borer *(Spodoptera litura)*
			2.0–2.5 kg	Cotton	Bollworm *(Heliothis armigera)*
			1.0–3.0kg	Rice	Leaf folder *(Cnaphalocrocis medinalis)*
2.	*Bacillus thuringiensis* var. *kurstaki* 5% WP	Bacterium	500–1000 gm	Cotton	American Bollworm
			750–100 gm		Spotted Bollworm
			1000–1250 gm	Red gram	Pod Borer
			500–1000 gm	Cabbage	Diamond back moth

3.	*Bacillus thuringiensis* var. *kurstaki, serotype h-39, 3b, strain z-52* Bio-tech. International	Bacterium	0.75–1.0 kg.	Cotton	Bollworms, Spodoptera
			1.50 kg.	Rice	Stem borer and Leaf folder
			0.75 kg.	Gram	*Heliothis*
			0.75 kg.	Pigeon pea	*Heliothis*
			0.75 kg.	Soybean	*Spodoptera, Heliothis, Spilosoma, Semilooper miner,* leaf
			1.50–2.00 kg.	Tobacco	*Spodoptera, Heliothis*
			1.00 kg.	Castor	Hairy caterpillar, *Achaea janata*
			0.25–0.50% Sol.	Teak	Defoliator *(Hyblaea puera)* Skeletonizer *(Eutectona machaeralis)*

MAJOR USES OF PESTICIDES
(Registered under the Insecticides Act, 1968)
UPTO- 31.08.2015

Sr no.	Microbial Insecticides	Texas ingredients	Dosage / ha Formulation (gm/ml)	Target Crops	Target Pest
1.	*Beauveria bassiana* 1.15% W.P.	Fungus	2 kg/ha	Cotton	Bollworm
			2.5 kg/ha	Rice	Leaf folder
2.	*Beauveria bassiana*-1% WP strain no: NBRI- 9947	Fungus	3.0 kg/ha	Chick pea	Pod borer
3.	*Beauveria bassiana* 10% SC	Fungus	1–1.5 kg/ ha	Cabbage	DBM
4.	*Beauveria bassiana*-1% International Panacea Ltd. STRAIN NO. IPL/BB/MI/01	Fungus	3.75–5.0 kg	Okra	Fruit borer / spotted bollworm
5.	*Verticillium lecanii* 1.15%WP	Fungus	2.5 kg	Cotton	White flies
				Citrus	Mealy bug

MAJOR USES OF PESTICIDES
(Registered under the Insecticides Act, 1968)
UPTO- 31.08.2015

Sr no.	Microbial Insecticides	Texas ingredients	Dosage / ha	Target	
			Formulation (gm/ml)	Crops	Pest
1.	Nuclear Polyhedrosis Virus of *Helicoverpa armigera* 0.43% AS	Virus	2700	Cotton	*Helicoverpa armigera*
			1500	Tomato	*Helicoverpa armigera*
2.	NPV OF *Helicoverpa armigera* 2.0% AS	Virus	250–500	Pigeon pea	Pod borer
				Chick pea	Pod borer
				Tomato	Fruit borer
				Gram	Pod borer
3.	NPV OF *Spodoptera litura* 0.5%AS	Virus	1500	Tobacco	*Spodoptera litura*

MAJOR USES OF PESTICIDES
(Registered under the Insecticides Act, 1968)
UPTO- 31.08.2015

Sr no.	Microbial Nematicides	Texas ingredients	Dosage / ha	Target	
			Formulation (gm/ml)	Crops	Pest
1	*Paecilomyces lilacinus* 1.15% WP	Fungus	3 kg/ha	Brinjal	Root Knot nematode
2.	*Verticillium chlamydosporium*	Fungus	–	Vegetables and cereal crops	Root Knot nematode

(Source: According to the recommendation of Govt. of India, Ministry of Agriculture-Major use of pesticide: Registered under the Insecticide Act, 1968. As on 31.08.2015).

Bio-pesticides Industry in India grown at a CAGR of 20.2% since 2010–2020

- Scope of Current market: USD 23.92 million (2015) for pesticides
- Represent only 4.2% of the overall pesticide market
- Over 150 bio-pesticide producing companies
- 15 types of bio-pesticides out of 227 pesticides registered

- Highest demand for bio-pesticides was observed from West India-Maharashtra followed by South India
- Hugely Driven by Microbial pesticides sale of *Trichoderma viride*, *Pseudomonas fluorescens* and *Bacillus thuringensis* (Courtesy: India Biopesticide Market Outlook: 2020 Ken Research Report, 2015.)

Registration and the Regulatory System

Bio-pesticides fall under the Insecticide Act (1968) under which any microbial organism manufactured or sold for pest and disease control should be registered with the Central Insecticides Board (CIB) of the Ministry of Agriculture. Manufacturers can register their products under either 9(3) B (temporary registration) or 9(3) (regular registration).

The data requirement for registration under 9(3) B is less stringent than for 9(3). For example, efficacy data on specified crops are required from 2 locations over two seasons for 9(3) B, while the same is required from 3 locations for 9(3). Data on product characterization, efficacy, safety, toxicology, and labelling must be submitted while applying for registration.

The CIB's established quality standards must be met, with reference to content, virulence of the organism in terms of LC_{50}, moisture content, shelf-life, and secondary non-pathogenic microbial load. Protocols for assessing these quality parameters have been prescribed (Rabindra, 2005). Surveys to test the quality of bio-pesticides are conducted, and while some manufacturers clearly meet accepted standards (Ignacimuthu *et al.,* 2001), other reports indicate quality concerns, especially from new and inexperienced producers.

Indian companies have formed a bio-pesticide supplier's association, the All India Biotech Association, to co-ordinate the commercial sector's voice in developing government policy. Other organizations actively promoting bio-pesticides include non-governmental organizations (NGOs), such as the M.S. Swami Nathan Research Foundation, and International Research Centres based in India like International Crops Research Institute for the Semi-Arid Tropics (ICRISAT) and the International Rice Research Institute.

References

Armes, N.J., D.R. Jadhav, G.S. Bond and A.B.S. King (1992). Insecticide resistance in *Helicoverpa armigera* in south India. *Pest. Sci.* 34: 355–364.

Biji, C.P., W. Suheendrakumar and T.V. Sajeev (2006). Quantitative estimation of production of *Hyblea puera* NPV in three stages of teak defoliator. *J. Virol. Methods* 136: 78–82.

Burges, H.D. (1981). Safety, safety testing and quality control of microbial pesticides. Pages 738–769. In: Microbial Control of Pests and Plant Diseases. H.D. Burges (ed.). Academic Press. London.

Burges, H.D., G. Croisier and J. Huber (1980). A review of safety tests on baculoviruses. *Entomophaga*, 25: 329–340.

Cook, R.J., W.L. Bruckart, J.R. Coulson, M.S. Goettel, R.A. Humber, R.D. Lumsden, J.V. Maddox, M.L. McManus, L. Moore and S.F. Meyer (1996). Safety of micro-organisms intended for pest and plant disease control: a framework for scientific evaluation. *Biol. Control* 7: 333–351.

Ignacimuthu, S. and A. Sen (Ed.) (2001). *Microbials in insect pest management.* Science Publishers Inc. Enfield, USA.

Jaronski, S.T., M.S. Goettel and C.J. Lomer (2003). Regulatory requirements for Eco-stoxicological assessments of microbial insecticides – how relevant are they. pp. 236–260, *In* H.M.T. Hokkanen and A.E. Hajek (ed.), *Environmental Impacts of Microbial Insecticides.* Kluwer Academic Publishers. Dordrecht, The Netherlands.

Kabaluk, J.T., Svircev, A.M., Goettel, M.S. and Woo, S.G. (eds.) (2010). *The use and regu-lation of microbial pesticides in representative jurisdictions worldwide.* International Organization for Biological Control of Noxious Animals and Plants (IOBC). pp. 108.

Kranthi, K.R., D.R. Jadav, S. Kranthi, D.R. Wajari, S.S. Ali and D.R. Russell (2002). Insecticide resistance in five major insect pests of cotton in India. *Crop Prot.* 21: 449–460.

Mensink, B.J.W.G. and J.W.A. Scheepmaker (2007). How to evaluate the environmental safety of microbial plant protection products: A proposal. *Biocontrol Sci. Technol.* 17: 3–20.

Organization for Economic Cooperation and Development (2003). Guidance for Registration Requirements for Microbial Pesticides. Environment Directorate. ENV/JM/MONO (2003)5. Available online at www.oecd.org/dataoecd/ 4/23/28888446.pdf.

Rabindra, R.J. (2005). Current status of production and use of microbial pesticides in India and the way forward. pp. 1–12, *In* Rabindra, R.J., S.S. Hussaini, and B. Ramanujam (ed.), *Microbial Biopesticde Formulations and Application.* Technical Document No. 55. Project Directorate of Biological Control.

Singhal, V. (2004). Bio-pesticides for sustainable agriculture: prospects and constraints. pp. 31–40, *In* E. Kaushik (ed.), *Biopsticides in India.* All India Biotech Association 'Vipps Centre'. New Delhi.

Veenakumari K., R.J. Rabindra, C.D. Srinivasa Naik and M.R. Shubha (2007). *In situ* field level mass production of *Amsacta albistriga* (Lepidoptera: Arctiidae) nucleo-polyhedrovirus in a groundnut ecosystem in South India. *Biol. Abstr. Int. J. Trop. Insect Sci.* 27: 48–52.

Wahab, S. (2004). The Department of Biotechnology initiates towards the development and use of bio pesticides in India. pp. 73–90, *In* E. Kaushik (ed.), *Bio-pesticides for sustainable agriculture: prospects and constraints.* New Delhi: The Energy and Resources Institute.

Question times

1. What are the characteristics of effective Regulatory Systems for pest control?
2. Discuss the Registration and the Regulatory System of Bio-pesticides in India.

One word answer

1. Fungus used for management of DBM in cabbage - ***Beauveria bassiana.***
2. Fungus used for management of Root Knot nematode - ***Paecilomyces lilacinus.***
3. Virus used for management of Gram Pod borer - **NPV of *Helicoverpa armigera.***
4. Bacterium used for management of Fruit borer of Chillies - ***Bacillus thuringiensis* var. *Galleriae.***
5. Fungus used for management of White flies in Cotton - ***Verticillium lecanii.***

■■■■

Chapter 8

USE OF INSECT PATHOGEN IN INTEGRATED PEST MANAGEMENT

Introduction

Phytophagous Insects cause a considerable loss to crops in the countries with tropical and subtropical climate like India. Major pests include the arthropods from various orders *viz.* Lepidoptera, Hemipteran, Diptera, Coleoptra, Isopteran and Hymenoptera along with 130–150 distinguished species of mites. For their control, most of the farmers use chemical pesticides which have adverse impact on environment and human health. In India, insecticides are used in a large amount than other pesticides, such as fungicides, acaricide, nematicide and herbicides. Conventional insecticides acquire inherent toxicities that jeopardize to the environment and also affect human beings and other living organisms. Negative effects of conventional insecticides on human health led to a resurgence in interest in microbial insecticides because of their minimal ecological side effects. To solve the problems mentioned above, there is a need of modern concept of pest management based upon an understanding of agroecosystem ecology, and integration of various control tactics into pest management systems.

IPM is a tactics of pest management including monitoring of insect populations, cultural, mechanical, biological and chemical methods of insect-pest's management. In all several IPM tactics, there is biological pests control technique that discourage the development of populations of harmful organisms. The manipulation of beneficial organisms or natural enemies remains a very important tool in integrated pest management program of insect-pests all over the world. In this chapter we have descried implementation of biological control tactics especially insect pathogen which acts useful in IPM. It also explains how biological control interacts with other IPM tactics. Biological control, *i.e.* natural enemies, has been accepted as an effective, environmentally non-degrading, technically appropriate, economically viable and socially acceptable method of pest management (Gangwar, 2017).

Field and laboratory evaluation experiments indicate that entomopathogenic microorganisms may be adequately used in the management of insect and mite pests in cereals, vegetables, fruit crops, flowers and forest crops. The highest promise probably lies with fungal pathogens (*Beauvaria bassiana, Hirsutella thompsonii, Metarhizium anisopliae, Nomuraea rileyi, Entomophthora thaxter-iana, and E. parvispora)*, but bacterial (*Bacillus thuringiensis*), microsporidian

(*Nosema locustae*) nematode (*Steinernema feltiae*) and even viral (*Baculoviruses*) (Odindo, 1992) pathogens may be exploited in an integrated pest management programme of sub-tropical agricultural, horticultural and silvicultural crop pests.

8.1. Approach to Biological Control

Biological control can be defined as the utilization of natural enemies to reduce the damage caused by noxious organisms to tolerable levels. Biological control agents have been utilized for the management of insect-pests since long time in the different parts of the world. However, in last 3–4 decades we have seen a dramatic increase in their use after the recognized some drawback of conventional pesticides against environment and human being. After this incident, farmers as well as policy makers created an interest to understand how they can be better manipulated as part of effective, safe, pest Management systems are revealing surprising complexities in the life histories of these organisms.

8.2. Essential Characteristics of Effective Bio-control Agents

- Speed/Mobility to prevent pathogen to develop resistant structures.
- Longevity, enough to protect plant during its vulnerable period.
- Environmental tolerance, to sustain activity under different soil and climatic conditions.
- Mode of Action varies from pathogen to pathogen, physical contact, chemical nature of killing component.

8.3. Inclusion of bio pesticide in IPM

IPM aims at suppressing the pest species by combining more than one method of pest control in a harmonious way with least emphasis on the use of insecticides. In simple terms "IPM is the right combination of cultural, biological and chemical measures which provide the most effective, environmentally sound and socially acceptable methods of managing diseases, pests and weeds". The IPM seems to be the only answer to counter some of the major pests of crops, which have become unmanageable in recent years. The success of IPM largely depends upon conservation of naturally occurring bio-control agents.

8.3.1 What is a Bio-pesticide?

Bio-pesticides are living organisms which can intervene the life cycle of insect pests in such a way that the crop damage is minimized. The agents employed as bio-pesticides, include parasites, predators and disease causing fungi, bacteria and viruses, which are the natural enemies of pests. Further, they complement and supplement other methods of pest control. Utilisation of naturally occurring parasites, predators and pathogens for pest control is a classical biological control. On the other hand, these bio-agents can be conserved, preserved and multiplied under laboratory condition for field release. Once these bio-agents are introduced in the field to build their population considerably, they are capable of bringing down the targeted pests' population below economic threshold level (ETL).

8.3.2 Working principles of Bio-pesticides

In nature every ecosystem exists in a balance. Growth and multiplication of each organism depends on the food-chain, its predators, parasites, *etc*. In biological control system, these interrelations are exploited. The natural enemy of a pest, disease or weed is selected; its biology is studied for mass multiplication and utilizes the same to check the target pest. They are also specific in their action and perish once their feed (*i.e.* the pest) is exhausted. Thus they are based on natural principles, do not leave any residue, safe and economical.

8.3.3 Major advantages of Bio-pesticides

Bio-pesticides are preferred over chemical pesticides for the following reasons:

1. **Target specific and safe to beneficial organisms like pollinators, predators, parasites *etc.*:** The organisms used in microbial insecticides are essentially nontoxic and non-pathogenic to wildlife, humans, and other organisms not closely related to the target pest. The safety offered by microbial insecticides is their greatest strength.
2. **Growth of natural enemies of pests is not affected, thus reducing the pesticide application:** The toxic action of microbial insecticides is often specific to a single group or species of insects, and this specificity means that most microbial insecticides do not directly affect beneficial insects (including predators or parasites of pests) in treated areas.
3. **Environmental friendly**: In some cases, the pathogenic microorganisms can become established in a pest population or its habitat and provide control during subsequent pest generations or seasons.
4. **No harmful residues:** Due to their residues presents no hazards to humans or other animals, microbial insecticides can be applied even when a crop is almost ready for harvest.
5. **Important component of IPM**: If necessary, most microbial insecticides can be used in conjunction with synthetic chemical insecticides because in most cases the microbial product is not deactivated or damaged by residues of conventional insecticides (Follow label directions concerning any limitations). So bio-pesticides can be used as 1st line and 2nd line of defence, chemicals being the last resort.

6. **Cost effective**: They also enhance the root and plant growth by way of encouraging the beneficial soil microflora. By this way they take a part in the increase of the crop yield.

8.4. Limitations or Disadvantages of Microbial Control

1. **High Selectivity/Host Specificity:** Since the microbial agent is effective against only single pest species, there is a very limited market potential.
2. **Requirement of Additional Control Measures:** An additional control measures are needed for non-target insect pest species. It presents and not kept below ETL by natural control factor.
3. **The Correct Time of Application:** In this respect, the incubation period of disease, which is a critical factor and difficult to judge because of age specificity and stage specificity. The pathogens attack on certain stage (larval) and do not produce effect on eggs or adults. Susceptibility of larvae also decreases with increase in age.
4. **Delayed Effect/Mortality:** The pathogens require some time to produce response in their target host owing to its long disease incubation period.
5. **Difficulties in Applications**: The pathogens like viruses, bacteria must be ingested as early as possible to produce effects, while the fungi required entry through body integument for which good coverage of spray deposits is important.
6. **Storage Problem:** It is necessary to maintain the pathogen in viable condition. Even average temperature and sunlight degrade the microbial preparations. Exposure to sunlight causes UV radiation and losses it's virulent immediately. The pathogens have very short self-life.
7. **Difficulty of Culturing in Large Quantities:** Rearing host insect on artificial diet is more cumbersome and laborious works which requires constant supervision. The whole organisms' techniques for *in vivo* prorogation of virus are fairly laborious and expensive.
8. **Short Residual Effectiveness:** This is the major limitation absent the effectiveness of pathogen under field conditions as sunlight causes UV radiation.
9. Not legally protected and data needed for registration is highly expensive
10. **Risk of generalization:** The use of Entomopathogenic insecticides should not be generalized because each pest has its own case.

They are specific to target insects and highly safe to mammals and the environment. Some fungi are effective microbial control agents against certain insect pests only under conditions of high R.H. and temperature that are available in glass houses. In some instances, fungi (*e.g., Beauveria bassiana*) cause allergy to man. They attack non-target insects, adult parsitoids and predators. Thus, fungi have low specificity and pose hazards to biodiversity as comparison to viruses and bacteria.

8.5. Status of Bio-pesticide use in India

Last decade has witnessed a tremendous breakthrough in this aspect, especially on standardization of production techniques of *Trichoderma, Gliocladium, Paecilomyces, Pseudomonas, Trichogramma,* NPV and *Bacillus* to use them against many insect pests and diseases. There are a number of instances where bio-control agents have been successfully employed in India. Some examples of these are given below:

- Growth of lantana weed was controlled by using the bug *Telonemia scrupulosa.*
- Sugarcane pyrilla has been successfully controlled in a number of States by the introduction of its natural enemy *Epiricania melanoleuca* and *Tetrastictus pyrillae*.
- *Trichogramma*, which feeds on the eggs of sugarcane borers, has been used against the borers in the states of Tamil Nadu, Rajasthan, UP, Bihar and Haryana.
- Similarly *Trichogramma, Bracon, Chelonus* and *Chrysopa* sp. are being used for the control of cotton bollworms.
- *Trichogramma* has also been used against rice stem borer and leaf folder.
- The sugarcane scale insect has been controlled with the help of predatory *coccinellid beetles* in UP, West Bengal, Gujarat and Karnataka.

8.6. Tools of Biological Control

Several living organisms are used for insect-pest's management as biological control agents depending upon availability of suitable formulation and feasibility of conditions for application. Biological control is often shortened to biocontrol. Examples of biocontrol include treatment of turf with spores of the bacterium *Bacillus popilliae*, which cause milky disease in Japanese beetle larvae.

Natural enemies of insect pests, also known as biological control agents, include predators, parasitoids, and pathogens. Biological control agents of plant diseases are most often referred to as antagonists. Biological control agents of weeds include seed predators, herbivores and plant pathogens.

However, in the remaining part of this chapter we will discuss about microbial biological control agents of insects *i.e. insect pathogens.*

8.6.1. Entomopathogenic Viruses

The first accounts of Baculovirus infections came from diseased silkworms in ancient China (Miller, 1997). It was also during this time that the large nuclear polyhedrosis viruses (NPVs) were distinguished from the much smaller granular looking viruses *i.e.* Granulosis viruses (GVs). "Bergold" was the first to characterize the rod-shaped virions found within the occlusion bodies in the 1930s and 1940s. It was also during this time that the utility of baculoviruses as biocontrol agents was described. Viruses have been isolated from more than a thousand species of insects from at least 13 different insect orders (Tanada and Kaya, 1993).

Entomopathogenic viruses from almost a dozen viral families have been isolated: Ascoviridae, Baculoviridae, Birnaviridae, Iridoviridae, Nodaviridae, Parvoviridae, Picornaviridae, Poxviridae, Reoviridae, Rhabdoviridae, and Tetraviridae (Adams and Bonami, 1991; Murphy, 1995). But the Baculoviruses *i.e.* family currently consists of two genera, the Nucleopolyhedroviruses (NPVs) and Granuloviruses (GVs). These new genera replace the Nuclear Polyhedrosis Viruses (Subgenus A) and Granulosis Viruses (Subgenus B), respectively.

A. General characteristics of entomopathogenic viruses

EPV are obligate intracellular parasites having either DNA or RNA encapsulated into a protein coat known as capsid to form the virions or nucleocapsids. The Nucleocapsids, when found embedded in a protein matrices, then they belong to the termed group of occluded viruses like those of NPVs, CPVs and GVs. Nucleocapsids that naturally lack such matrices termed as free viruses FVs.

The occluded viruses of the families Baculoviridae, Reoviridae, and Entomopoxviridae could withstand abiotic factors of the surrounding environment. Meanwhile, free virus members of the Rhabdoviridae that are associated with their host all the time avoid such environmental factors being vertically transmitted from generation to another (Tanada and Kaya, 1993). Otherwise, the rhabdovirus degrades in less than a week if left unprotected in the environment (Hochberg and Waage, 1991).

B. Effectiveness of Entomopathogenic Viruses

The effectiveness of these viruses as microbial control agents depends on their persistence in nature and the mode of dispersal in the surrounding environment (Seufi and Osman, 2005).

Case study 1 - The unique successful control of the coconut palm rhinoceros beetle, *Oryctes rhinoceros* with the virus, *Rhabdionvirus oryctes* discovered by Hüger (Hüger, 1966) in Malaysia is attributed to the vertical transmission of the virus and its teratological effects (Monty, 1974).

Case study 2 - In USA and Canada, the sawfly, *Diprion (Gilpinia) hercyniae*, was introduced from Europe and became a dominant forest pest. By all means including use of the common chemical insecticides, the control of this pest was not successful. The microbial control with the virus, *Borrelinavirus diprion* was the only effective measurement that suppressed the pest population. For this aerial applications of NPV-suspensions were carried out on large scales over the infected forests (El-Husseini, 2006).

C. Mode of action of Entomopathogenic Viruses

- Typically, the initial infection occurs when a susceptible host insect feeds on plants that are contaminated with the occluded form of the virus.
- When the OBs are ingested by the insect, the protein matrix dissolves in the alkaline environment of the host midgut (pH 8.0), releasing the infective particles (virions or ODV/PDV) into the midgut.
- The virus replicates within the nuclei of susceptible tissue cells.
- Tissue susceptibility varies greatly between viruses with some NPVs being capable of infecting almost all tissue types and most GVs being tissue-specific replications (*e.g.* fat body cell only).
- The BV initiates infection to other tissues in the haemolymph, *e.g.* fat bodies, nerve cells, haemocytes etc.
- The cells infected in the second round of virus replication in the insect larva also produce BV, but in addition, occlude virus particles within polyhedra in the nucleus.
- The accumulation of polyhedra within the insect proceeds until the host consists almost entirely of a bag of virus.
- In the terminal stages of infection, the insect liquefies and thus releases polyhedra, which can infect other insects upon ingestion.
- A single caterpillar at its death may contain over 10^9 OBs from an initial dose of 1000.
- The infected larvae exhibit negative geotropism before succumbing to the virus infection, thereby facilitating widespread dissemination. The speed with which death occurs is determined in part by the environmental conditions.
- The pest may be killed in 3–7 days in optimal condition, but may take 3–4 weeks in un-ideal condition (Cunningham, 1995; Flexner and Belnavis, 2000).

D. Control of pest insects with *viral* pesticide products

- **In India,** many new suppliers of HearNPV and SpltNPV have appeared in recent years following the adoption of simplified microbial pesticide registration and in response to the growing problem of synthetic insecticide resistance (Department of Biotechnology India, 2007; Rabindra and Grzywacz, 2010).

- **Australian growers** have incorporated BV for management of *H. armigera* in field crops, and importation of *Helicoverpa zea* NPV (HezeSNPV) for *H. armigera* control is now supplemented by local sources of a HearSNPV isolate (Buerger *et al.,* 2007; Hauxwell *et al.,* 2010)
- The development of some **forest pest** BV, such as the gypsy moth NPV has continued and commercial production of sawfly *Neodiprion abietis* is going on (Lucarotti *et al.,* 2007)
- **Turfgrass** pest control has also been a focus for pests such as *Agrotis ipsilon* using an NPV (AgipMNPV, Prater *et al.,* 2006).

8.6.2. Entomopathogenic Bacteria

The discovery of *Bacillus thuringiensis* by Dr. Ernest Berliner in Germany in 1915 (Berliner, 1915) drew the attention of Dr. Edward Steinhaus in California, USA, during the late 50s, *B. thuringiesis* (Bt.) was used on large scale applications successfully to control the larvae of the Luzerne moth, *Goliath philodice* in California (Steinhaus, 1963).

A natural occurring Bt. strain may require genetic improvement before it can be formulated into an effective bio-pesticide. Any Bt. strain can be improved to increase toxicity and the range of target pests and to delay the onset of pest-resistance by including toxins that bind to different sites or have different modes of action. It is generally assumed that total bioactivity of a Bt strain is a function of additive and/or synergistic interactions of individual Cry proteins present in their proportional amounts, hence strain improvement has been attempted by increasing the copy number and type of Cry genes in a strain,

A. General characteristics of entomopathogenic bacteria

Entomopathogenic bacteria are unicellular prokaryotic organisms having size ranging from less than 1 μm to several μm in length. Bacteria with rigid cell walls are cocci, rod-shaped and spiral, while bacteria without cell walls are pleomorphic.

The majority of bacterial pathogens of insect-pests occur in bacterial families Bacillaceae, Pseudomonadaceae, Enterobacteriaceae, Streptococcaceae, and Micrococcaceae. These families of bacteria usually represent epiphytes or weak pathogens; however, some of them are highly virulent to their respective hosts.

Members of this entomopathogenic group of bacteria can be found in most ecological niches. In natural habitats, several Bt. isolates have no known target, as opposed to early Bt. isolates, which were known to be pathogenic for insects. This lack of insecticidal activity may be attributed to the loss of ability to produce insecticidal crystalline proteins (ICPs). Or simply, a test insect-pest for the actual target of that isolate is as yet unknown.

B. Effectiveness of entomopathogenic bacteria

Based on safety to non-target organisms and humans, only the members of Bacillaceae (Order Eubacteriales) were the most studied, commercialized, and successfully used in microbial control of lepidopteran, dipteran and coleopteran insect pests.

Case study 1 - The first microbial control treatment using an entomopathogenic bacterium was that against the grubs (larvae) of the Japanese beetle, *Popillia japonica*. The Bacterium *Bacillus popilliae*, the causative agent of the milky disease was produced by injection into grubs collected from the fields. Processed formulations were applied in the field or turned into soil. The disease suppressed the pest population for more than 10 years after one application. The dead grubs became an infection source in soil for other healthy individuals. In fact, this unique successful example could be attributed to the presence of *B. popilliae* spores in soil far from the destructive effect of the sun rays (UV).

Case study 2 - Another subspecies of Bt, *B. thuringiensis japonensis* strain *Buibui* has good potential for suppressing white grubs, especially Japanese beetle *P. japonica* (Bixby *et al.*, 2007), but it is not yet registered for commercialization.

C. Mode of action of entomopathogenic bacteria

- Bt. produces a para-sporal inclusion, a protein crystal body during sporulation.
- Many distinct crystal protein (Cry) genes have been described.
- The gut epithelium is the primary target tissue for Bt. delta endotoxin action.
- Upon ingestion, the crystals dissolve in the alkaline environment of the midgut and then the protoxin is proteolytically processed to produce the actual toxin.
- Activation of actual toxin usually involves the removal of a small number of N-terminal amino acid residues along with the cleavage of the C-terminal half (Gill *et al.*, 1992).
- The activated toxin then binds to specific receptors present on the membranes of the host insect epithelial midgut cells and induces the formation of pores in the membrane of midgut epithelial cells.
- This is followed by an increase in cell membrane permeability which eventually leads to cell lysis, disruption of gut integrity and finally to the death of the insect from starvation or septicaemia (Adang 1985; Gill *et al.*, 1992; Bauer, 1995).

- In short we can say that the crystal proteins exert their effect on the host by causing lysis of midgut epithelial cells, which leads to gut paralysis. The insect stops feeding and, if it does not recover, eventually dies.

D. Control of pest insects with *B. thuringiensis* microbial pesticide products

- **Crops and orchards.** Bt. has no pre-harvest spray interval and can be applied until harvest begins. It has minimal or no impact on beneficial organisms in these agroecosystems; however, although efficacious, it is sensitive to solar degradation and requires frequent application.
- It is used extensively in organic vegetable production and is increasingly being utilized by conventional growers. Control of a plethora of pest Lepidoptera is common in row crops including crucifers, solanaceous vegetables, cucurbits, corn, legumes, soybeans, cotton, and others.
- The implementation of Btk for control of orchard pests, particularly leaf rollers and other defoliators, was described by Lacey *et al.* (2007)
- **Stored product pests.** Several pest insects attack stored grain, fruit, nuts, potatoes and other stored food products.
- Good efficacy of Btk has been demonstrated and protocols have been published for the evaluation of Btk control against *Plodia interpunctella* and other lepidopteran pests of stored grain (Lord et al., 2007).
- Lacey and Kroschel (2009) described examples of large-scale implementation of Btk in several countries for control of the potato tuber moth, *P. operculella* in rustic stores of potato tubers.
- **Lawn and turf.** Koppenhöfer *et al.* (2012) provide overviews of the use of Bt. for control of turf pests. Btk and Bt. subsp. *aizawai* are registered for control of sod webworms and armyworm, *Mythimna (Pseudaletia) unipuncta*, in turf.

8.6.3. Entomopathogenic Fungi

Entomopathogenic fungi is identified as a promising bio-control agent in the regulation of insect pest population without harming the non-target insects. Over 800 species of entomopathogenic fungi have been described and identified. Most research on entomopathogenic fungi has been aimed at developing them as inundative biological control agents of insects, mites and ticks, despite the great potential for use in conservation and classical biocontrol strategies.

In 1980s, the first insect pathogenic studies were carried out and their focus was to find the methods of disease management of the silkworm. Bassi in 1835, first time formulated the germ theory by the use of white muscardine fungus on the silkworm that was then named in his honor as *Beauveria bassiana.* Gilbert and Gill (2010) described that this silkworm disease gave the idea of using insect infecting fungi for the control of insect pest management.

A. General characteristics of entomopathogenic fungi

These fungi usually attach to the external body surface of insects in the form of microscopic spores (usually asexual, mitosporic spores also called conidia). Under the right conditions of temperature and (usually high) humidity, these spores germinate, grow as hyphae and colonize the insect's cuticle; which they bore through by way of enzymatic hydrolysis reaching the insects' body cavity or haemocoel (Fernandes *et al.,* 2012). Then, the fungal cells proliferate in the host body cavity, usually as walled hyphae or in the form of wall-less protoplasts (depending on the fungus involved). After some time the insect is usually killed (sometimes by fungal toxins) and new propagules (spores) are formed in or on the insect if environmental conditions are again right. High humidity is usually required for sporulation.

B. Effectiveness of entomopathogenic fungi

There are more than 400 species recorded as entomopathogenic fungi from which only about 20 species with potential capability for use in microbial control of insect pests (Zimmerman, 1986). Most of them are included in 11 genera, *i.e.,* Lagnidium, Entomophthora, Neozygites, Erynia, Ashersonia, Verticillium, Nomuraea, Hirsutella, Metarhizium, Beauveria and Paecilomyces.

The early successful application with an entomopathogenic fungus dated back to 1878 by the Russian scientist Metchnikoff using *Metrhizium anisopliae* against the wheat leaf beetle, *Anisoplia austriaca* and the beet weevil *Cleonus punctiventris* in sugar beet fields (Metchnikoff, 1879).

Case study 1 - The successful insect management is reported with certain fungi (*Verticillium lecanii, Beaveria bassiana*, and *Paecilomyces fumosoroseus*) used in glass houses against aphids, depending on the inside dominating high temperature and relative humidity, *e.g.,* in tropical and subtropical countries.

Case study 2 - The use of *B. bassiana* and *M. anisopliae* in Brazil was very effective for controlling the cacao coleopteran and lepidopteran insect pests.

C. Mode of action of entomopathogenic fungi

- The mode of action of insect pathogenic fungi varies and kills the insect by different ways such as causing starvation to toxin production.
- These insect pathogenic fungi produce many toxins and extracellular enzymes such as proteases and chitinases which aid penetration of the host physical defences.

- The infective unit in most fungi is a spore, usually a conidium. In many cases, the conidia are adhesive to the cuticle, or secrete adhesive mucus as the conidium swells during pregermination.

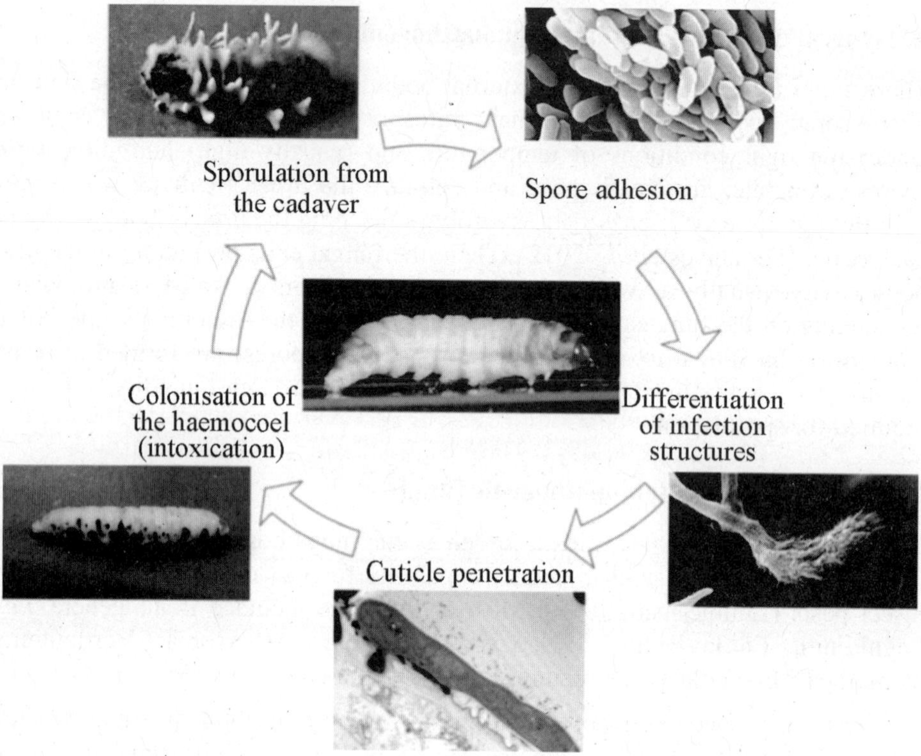

Figure 8.1. General overview of the invasive and developmental processes of entomopathogenic fungi in an insect host. The key stages include adhesion of infection propagules (*e.g.,* conidium and blastospore) to the surface of the host cuticle, differentiation of infection structures (appressoria and penetration pegs), penetration of the cuticle using enzymes and mechanical force, colonization of the hemocoel and emergence of the conidiophores for external sporulation on the cadaver. Picture source: www.researchgate.net/publication/295909185_Entomopathogenic_ Fungi_New_ Insights_into_Host-Pathogen_Interactions/figures?lo=1

- In favourable conditions, the conidium, germinates into a short germ tube which gives out small swellings called appressoria.
- The appressorium attaches itself to the cuticle and sends out an infection peg which provides a firm attachment that the fungus needs to physically force its way into the host. The hyphae then penetrate the layers of the integument by enzymatic dissolution of chitin and protein, ramify first in the cuticle and then reach the haemocoel and internal organs.
- The invasion by the fungal mycelium continues until the insect is virtually filled with the fungus and becomes quite firm to touch.

- Conidiophores are then produced which erupt through the cuticle and produce spores on the outside of the insect, infecting nearby healthy insects also.
- Death of the host is by obliteration (choking) of the tissues and also by the toxins produced by the fungus.

D. Control of pest insects with entomopathogenic fungi pesticide products

EPF are a major component of integrated pest management techniques as biological control agents against insect pests and other arthropods and are an integral part of myco-insecticides in horticulture, forestry and agriculture. Some of the commonly used EPF are as follows.

- ***Beauveria bassiana*** – These enter the host insects body through food or in contact with the host cuticle and reproduce inside the insect body. It produces toxins namely *beauvericin, bassianocideetc*. Inside the host body causes paralysis of the host insects and ultimately kills the insects within four or five days. They are used particularly to control sucking pests and caterpillars infesting crop plants. These fungi are used to control the caterpillars of yellow stem borer and leaf folder of rice, white grub of groundnut, sugarcane pyrilla, coconut rhinoceros beetle, caterpillars of pulses, tomato and cotton, diamond back moth, leaf eating caterpillars of tobacco and sunflower *etc*.
- ***Lecanicelium lecanii*** – This beneficial fungus is mainly used to control whiteflies, aphids, thrips, brown plant hopper, scale insects, mealy bugs and other sucking insect pests of crop plants.
- ***Metarhizium anisopliae***– This pathogenic fungus is used to control mainly coconut rhinoceros beetle, groundnut cut worm, rice brown plant hopper, diamond back moth and early shoot borer, top shoot borer and internode borer of sugarcane.
- ***Nomuraea rileyi*** – It is used to control pod borers, cut worms, cabbage borers etc.
- ***Hirsutella thompsonii***– These fungi are used to control different hoppers and bug pests, whiteflies, red mites etc.
- ***Paecelomyces fumosoroseus*** – This fungus is used to control yellow and red mite, whiteflies etc.

8.7. Safety Issue Related to Insect Pathogen used in IPM

Safety of entomopathogenic microorganisms used in microbial control is a major concern associated with the increased needs for biocontrol agents desired in the exploitation of IPM programs for agricultural, veterinary, and medical insect pests (Burges, 1981). In fact, most of the microbes used as biological control agents occur naturally in the different ecosystems (Dirimanov *et al.*, 1980), even in large densities when causing epizootics. It is worth to know that the general absence for infection of man with such Microbes in the medical literature is an important evidence that these agents do not pose a significant human health risk (Van Driesche and Bellows, 1996), or for plants (Campbell and Sands, 1992), fish and crustaceans (Spacie, 1992), birds (Kervin *et al.,* 1990), mammals (Siegel and Shadduck, 1992), and non-target insects and mites (Fisher and Briggs, 1992). The exception is that fungi show less specificity posing risk to beneficial like predatory insects, parasitoids, bees and pollinators. The level of specificity required for safety depends on the characteristics of both the targeted area and the taxonomic groups of involved organisms.

8.7.1. Safety of bacterial insecticides

In the very early stage of using *B. thuringiensis* in the 50s, there was a concern for its toxicity to mammals and non-target organisms due to the presence of the exotoxins produced by the vegetative cells either in the production medium or in the body of infected lepidopteran larvae remain in the ecosystem. Thus, some concern has been raised in the past about potential Bt. contamination of honeybee and of drinking water supplies by exotoxins in Germany due to its close relation to *B. cereus*, a bacterium implicated in food poisoning in humans. But the picture is completely different in the last 3 decades, since the discovery of the isolates that do not produce exotoxins which are currently the base for commercial production of Bt. (mostly Bt. subsp. *kurstaki*) in different countries.

The literature is very rich with results proving the safety of Bt. and *B. sphaericus* to mammals and non-target organisms. The infected lepidopteran prey could affect the predatory insects (El-Hussini *et al.,* 1985). For example, the silkworm, *Bombyx mori*, and immature parasitoids inside target pests are likely to be killed if exposed to Bt. This explains the prohibited use of Bt. as microbial pest control agent in India at the areas of silk production is an understandable special case to avoid infection of larval colonies mass reared on the mulberry trees in open fields.

Also, when Bt. subsp. *israelensis* was applied to aquatic systems, it showed little effect on non-target invertebrates, with the exceptions of those in the dipteran families Chironomidae, Dixidae, and Ceratopognidae that were killed. Some effects were also recorded on Ephemeroptera and Odonata, but much less than that caused by application of chemical pesticides in the same circumstances. It could be related in case of Bt.subsp. *israelensis* to suppression of mosquito larval population as prey for immatures of Odonata (Zgomba *et al.,* 1986).

8.7.2. Safety of fungal insecticides

Most of the entomopathogenic fungi developed for commercial use in microbial control of insect pests showed no infectivity to man or other vertebrates (Podgwaite, 1986). Safety tests with *Nomuraea rileyi,* (Ignoffo, 1973; Ignoffo *et al.*, 1979) *Hirsutella thompsonii,* (McCoy and Heimpel, 1980) *Verticillium lecnii*

(Podgwaite, 1986) *and Lagenidium giganteum* (Kervin *et al.*, 1990) assured negative findings to different mammalians and birds (Kervin, 1992).

On the other hand, *Beauveria bassiana* has been reported to cause allergies in humans and is at least an opportunistic pathogen to man and other mammals (Burges, 1981). Concerning non-target invertebrates, high mortality appear when they contacted or ingested spores of the entomopathogenic fungi. Larvae of the coccinellid *Cryptolaemus montrouzieri* suffered 50 per cent mortality when fed Boverin-t, a commercial conidiospore preparation of *Beauveria bassiana* (Flexner *et al.*, 1986). Honey bee workers experienced 29 per cent mortality when fed spores of *Hirsutella thompsonii* (Cantwell and Lehnert, 1979).

Both *B. bassiana and Metarhizium anisopliae* infect *B.mori*, and also killed honey bees following field applications (Podgwaite, 1986). Besides, the parasitized hosts of some species showed increased susceptibility to entomopathogenic fungi and those populations of some over wintering predacious carabid beetles and other invertebrates are killed by fungi showing an increased risk if large amounts of fungal inoculums are added to soils as a consequence of agricultural microbial pest control. (Flexner *et al.*, 1986). Infection of different beneficial natural enemies were recorded, (El-Husseini, 2006) *e.g.,* Adults of the braconid parasitoid *Apanteles* sp., the coccinellid predator *Cydonia vicina isis*, the earwig *Labidura riparia,* and the syrphid fly *Syrphus corollae* in fields of sugar beet treated with conidiospores of both *B. bassiana* and *M. anisopliae*.

8.7.3. Safety of viral insecticides

A literature summary concerned with safety of entomopathogenic viruses to human and environment was published (Summers *et al.,*1975; Van Driesche and Bellows, 1996). It is reported that the majority of viruses which have been considered for use in pest control are either NPVs or GVs of the family Baculoviridae that are specific to arthropods. Their safety was tested against more than 24 vertebrate species including different species of mammalian, avian, and fish. None of these viruses was able to cross infection to any of the tested animals (Podgwaite, 1986).

Concerning safety to beneficial arthropods, immature larvae of parasitoids in infected hosts may die not due to virus infection, but relatively to premature loss of the host or to alteration in quality of the host (Flexner *et al.*, 1986). Entomopathogenic viruses are typically limited in their host range, such that only species in one genus or in related genera in one family are infected. Thus, not only the distantly related types of invertebrates but also vertebrates are not at risk from virus applications (Flexner *et al.*, 1986). The NPV *Autographa californica* with the widest known host range (34 lepidopteran species) proved safe to beneficial and environment. Similar results were recorded for the CPV commercialized for use in Japan; one of them has a host range broader than most NPVs. The NPV infecting the lasiocampid larvae *Dendrolimus spectabilis.* Butler infects larvae in several genera of Lepidoptera (Van Driesche and Bellows, 1996).

8.8. Conclusions and Future Direction

Eco-friendly pest and disease management practices are being developed and evaluated globally to reduce the health risks due to higher usage of chemical pesticides in agriculture. In light of this, entomopathogenic bacteria and viruses have wide scope as biocontrol agents in addition to being a very good source for scouting for insecticidal toxin genes. An important benefit of microbial control agents is that they can be used to replace, at least in part, some more hazardous chemical pest control agents. The selective toxicity of these entomopathogenic bacteria and viruses to major insect-pests and their safety to non-target organisms makes them ideal tools for use in integrated pest management (IPM) programmes. These positive trends, however, need to be accompanied by strengthening of research efforts to overcome some of the major limitations in production, use and efficacy of baculoviruses. The relatively slow speed with which baculoviruses kill their hosts has hampered their effectiveness as well as acceptance by potential users. However, genetic improvement, using traditional methods as well as genetic engineering, may produce strains of baculoviruses with improved pathogenesis and virulence. Recent advances in virus production using insect cell lines offer a way out of this situation.

The future of microbial insecticides appears to be assured as a consequence of pest-resistance and environmental contamination with conventional insecticides. Also, advances in biotechnology should allow the production costs of bio-pesticides to decrease, in addition to increasing their efficacy. However, there is a pressing need to develop better formulations to enhance their efficacy as bio-pesticides in the field, although attempts have been made to genetically transform *P. fluorescens* with Bt. toxin gene to gain better field performance. Many Bt. subspecies have been registered, while many others have been described, but not developed commercially so far. Hence, there is a lot of scope for identification of novel toxin genes from these strains and to further develop them into bio-pesticides.

References

Adams J.R. and Bonami J.R. (eds.) (1991). Atlas of Invertebrate Viruses, CRC Press, Boca Raton, FL.

Adang, M.J., Staver, M.J., Rocheleau, T.A., Leighton, J., Barker, R.F. and Thompson, D.V. (1985). Characterised full length and truncated plasmid clones of the crystal protein of *Bacillus thuringiensis* subsp. *kurstaki* HD-73 and their toxicity to *Manduca sexta*. Gene 36: 289–300

Bauer, L.S. (1995). Resistance: a threat to the insecticidal crystal proteins of *Bacillus thuringiensis. Fla. Entomol.* 78: 414–442

Berliner, E. (1915). Über die Schlaffsucht der *Ephestiakühniella* und *Bacillus thuringiensis* n.sp. *Zeitschrift fürallgemeine Entomologie* 2: 21–56.

Bixby, A., Alm, S.R., Power, K., Grewal, P. and Swier, S. (2007). Suscepbility of four species of turfgrass-infesting scarabs (Coleopertra: Scarabaeidae) to *Bacillus thuringiensis serovar.japonensis* strain *Buibui. J. Econ. Entomol.* 100: 1604–1610

Buerger, P., Hauxwell, C. & Murray, D. (2007). Nucleopolyhedrovirus introduction in Australia. *Virologica Sinica,* 22(2): 173–179.

Burges, H.D. (1981). Safety, safety testing and quality control of microbial pesticides. Pages 738–769. In: Microbial Control of Pests and Plant Diseases. H.D. Burges (ed.). Academic Press. London.

Campbell, C.I. and D.C. Sands (1992). Testing the effects of microbial agents on plants. Pages 689–705. In: Microbial ecology: Principles, Methods and Applications. M.A. Levin, R.J. Seidler and M. Rogul (eds.). McGraw-Hill, New York, USA.

Cantwell, G.E. and T. Lehnert (1979). Lack of effect of certain microbial insecticides on the honeybee. *Journal of Invertebrate Pathology*, 33: 381–382.

Cunningham, J.C. (1995). Baculoviruses as microbial insecticides. In: Reuveni, R. (Ed.), Novel Approaches to Integrated Pest Management. CRC Press, Inc., Boca Raton, FL, pp. 261–292.

Dirimanov, M., R. Angelova and T. Sabrikova (1980). State of the harmful entomo- and acarofauna and predaceious species of insects in apple orchards with some plant protection technologies. Nauchni Trudova, *Entomology and Microbial Fitopathology*, 25: 15–30.

El-Husseini, M.M. (2006). Microbial Control of Insect Pests: is it an effective and environmentally safe alternative. *Arab Journal of Plant Protection*, *24*(2): 162–169.

El-Hussini, M.M., M.F.S. Tafik, K.T. Awadallah and A.I. Afifil (1985). Effect of *Bacillus thuringiensis* diseased prey on the development and activity of the anthocorid *Xylocoris flavipes* (Reuter). *Bulletin de laSocieté entomologique de Egypté*, 65: 205–215.

Fernandes, E.G., Valério, H.M., Feltrin, T. and Van Der Sand, S.T. (2012). "Variability in the production of extracellular enzymes by entomopathogenic fungi grown on different substrates". *Braz. J. Microbiol.* 43(2): 827–33.

Fisher, S.W. and J.D. Briggs (1992). Testing of microbial pest control agents in non-target insects and acari. Pages 761-777. In: Microbial ecology: Principles, Methods and Applications. M.A. Levin, R.J. Seidler and M. Rogul (eds.). McGraw-Hill, New York, USA.

Flexner, J.L., B. Lighthart and B.A. Croft (1986). The effects of microbial pesticides on non-target beneficial arthropods. *Agriculture, Ecosystem and Environment,* 16: 203–254

Flexner, J.L. and Belnavis, D.L. (2000). Microbial insecticides. In: Rechcigl, J.E., Rechigl, N.A. (Eds.), Biological and Biotechnological Control of Insect Pests. Lewis Publishers, Boca Raton, FL, pp. 35–61.

Gangwar, R.K. (2017). Role of Biological Control Agents in Integrated Pest Management Approaches. *Acta Scientific Agriculture*, 1: 09–11.

Gill, S.S., Cowles, E.A. and Pietrantonio, P.V. (1992). The mode of action of *Bacillus thuringiensis* δ-endotoxins. *Annu. Rev. Entomol.* 37: 615–636.

Gilbert, L.I. and Gill, S.S. (Eds.) (2010). *Insect Control : Biological and Syntehtic agents.* Academic Press.

Hauxwell, C., Tichon, M., Buerger, P. and Anderson, S. (2010). Australia. In: Kabaluk, J.T., Svircev, A.M., Goettel, M.S. and Woo, S.G. (eds.). *The Use and Regulation of Microbial Pesticides in Representative Jurisdictions Worldwide,* pp. 80–88. IOBC Global.

Hochberg, M.E. and J.K. Waage (1991). A model for the biological control of Oryctes rhinoceros (Coleoptera: Scrabaeidae) by means of pathogens. *Journal of Applied Ecology,* 28: 514–531.

Hüger, A.M. (1966). A virus disease of the Indian rhinoceros beetle, *Oryctes rhinoceros* (Linnaeus), caused by a new type of insect virus, *Rhabdionvirus oryctes* gen. N., sp. N. *Journal of Invertebrate Pathology*, 8: 3851

Ignoffo, C.M. (1973). Effects of entomopathogens on invertebrates. Annual New York Academy of Science, 217: 141–164.

Ignoffo, C.M., C. Garcia, R.W. Kapp and W.B. Coate (1979). An evaluation to the risks to mammals of the use of an entomopathogenic fungus, *Nomuraea rileyi*, as a microbial insecticide. *Environmental Entomology*, 8: 354–359.

Lacey, L.A., Arthurs, S.P., Knight, A.L. and Huber, J. (2007). Microbial control of lepidopteran pests of apple orchards. In *Field Manual of techniques in invertebrate pathology* (pp. 528–546). Springer, Dordrecht.

Lacey, L.A. and Kroschel, J. (2009). Microbial control of the potato tuber moth (Lepidoptera: Gelechiidae). *Fruit Veg. Cereal Sci. Biotechnol.* 3: 46–54.

Lord, J.C., Campbell, J.F., Sedlacek, J.D. and Vail, P.V. (2007). Application and evaluation of entomopathogens for managing insects in stored products. In *Field Manual of Techniques in Invertebrate Pathology* (pp. 677–693). Springer, Dordrecht.

Lucarotti, C.J., Moreau, G. and Kettela, E.G. (2007). A viral bio-pesticide for Control of the Balsam Fir Sawfly. *Biological Control: A Global Perspective: Case Studies from Around the World,* 353.

Kervin, J.L. (1992). Testing the effect of microorganisms on birds. Pages 729–744. In: Microbial ecology: Principles, Methods and Applications. M.A. Levin, R.J. Seidler and M. Rogul (eds.). McGraw-Hill, New York, USA.

Kervin, J.L., D.A. Dritz and R.K. Wahino (1990). Confirmation of the safety of *Lagenidium giganteum* (Oomycetes: Lagenidiales) to mammals. *Journal of Economic Entomology*, 83: 374–376.

Koppenhofer, A.M., Jackson, T.A. and Klein, M.G. (2012). Bacteria for use against soil-inhabiting insects. *Manual of techniques in invertebrate pathology,* 129–149.

McCoy, C.W. and A.M. Heimpel (1980). Safety of the potential mucoacaricide, *Hirsutella thompsonii*, to vertebrates. *Environmental Entomology*, 9: 47–49.

Metchnikoff, E. (1879). Zur Lehre über Insektenkrankheiten. *Zoologische Anzeiger*, 3: 44–47.

Miller L.K. (1997). "Introduction to the baculoviruses". The Baculoviruses, L. K. Miller, ed., Plenum Press, New York: 1–6.

Monty, J. (1974). Teratological effects of the virus *Rhabdionvirus oryctes* on *Oructes rhinoceros* (L.) (Coleoptera: Scarabaeidae). *Bulletin of Entomological Research,* 64: 633–636.

Murphy F.A. (1995). *"Virus Taxonomy: Classification and Nomenclature of Viruses".* Springer Verlag, Wien, New York.

Odindo, M.O. (1992). Future prospects for application of insect pathogens as a component of integrated pest management in tropical root crops. *Biocontrol Science and Technology*, *2*(3): 179–191.

Podgwaite, J.D. (1986). Effect of insect pathogens on the environment. Pages 279–187. International Symposium of the AKademie der Wissenschaften und der Literature, Mains, November 15–17, 1984, at Mains & Darmstadt., Fortschritte der Zoologie, (Volume 32, pp. 341), Gustav Fisher Verlag, Stuttgart, Germany.

Rabindra, R.J. and Grzywacz, D. (2010). India. In : Kabaluk, J.T., Svircev, A.M., Goettel, M.S. and Woo, S.G. (eds.). The use and regulation of Microbial Pesticides in representative jurisdictions worldwide. IDBC Global, p. 99.

Seufi, A.M. and G.E. Osman (2005). Comparative susceptibility of the Egyptian cotton leafworm, *Spodoptera littoralis* (Boisd.) to some baculovirus isolates Egyptian. *Egyptian Journal of Biological Pest Control*, 15(1&2): 21–26.

Siegel, J.P. and J.A. Shadduck (1992). Testing the effect of microbial pest control agents on mammals. Pages 745–759. In: Microbial ecology: Principles, Methods and Applications. M.A. Levin, R.J. Seidler and M. Rogul (eds.). McGraw-Hill, New York, USA.

Spacie, A. (1992). Testing the effects of microbial agents on fish and crustaceaus. Pages 707–728. In: Microbialecology: Principles, Methods and Applications. M.A. Levin, R.J. Seidler and M. Rogul (eds.). McGraw-Hill, New York, USA.

Steinhaus, E. (1963). Insect Pathology. An advanced treatise. Academic Press, New York & London. 455 pp.

Summers, M., R. Engler, L.A. Falcon and P. Vail (1975). Baculovirus for Insect Pest Control: Safety considerations. American Society of Microbiology, Washington DC.

Tanada, Y. and Kaya, H.K. (1993). Insect Pathology. Academic Press, Inc., San Diego, California, USA. 666 pp.

Van Driesche, R.G. and T.S. Bellows (1996). Biological Control. Chapman & Hall Publisher, USA. 539 pp.

www.agriinfo.in/default.aspx?page=topic&superid=6&topicid=787

www.bugsforbugs.com.au/images/Heliothis.JPG

www.geocities.com

www.insectariumvirtual.com

www.nabard.org

www.treknature.com

Zimmerman, G. (1986). Insect pathogenic fungi as pest control agents. Pages 217–231. In: Biological Plant and Health Protection. J.M. Franz (ed.). International Symposium of the AKademie der Wissenschaften undder Literature, Mains, November 15–17, 1984, Mains & Darmstadt., Fortschritte der Zoologie (Volume 32, 341 pp.) Gustav Fisher Verlag, Stuttgart, Germany.

Zgomba, M., D. Petrovic and Z. Srdik. (1986). Mosquito larvicide impact on May flies (Ephemeroptera) and dragon flies (Odonata) in aquatic biotypes. Proceedings of 3rd European Congress of Entomology, Amsterdam, 3: 532.

Question time

1. What is a bio pesticide? Discuss in brief the Working principles of Bio-pesticides?
2. What are the essential characteristics of effective Bio control agents?
3. Discuss the advantages of bio pesticides?
4. Describe the Limitations or Disadvantages of Microbial Control?
5. Discuss the Status of Biopesticide use in India with appropriate examples?
6. Illustrate the Mode of action of Entomopathogenic fungi with suitable diagram?
7. Discuss in brief the General characteristics of entomopathogenic fungi?
8. Illustrate the Mode of action of Entomopathogenic Bacteria?
9. Illustrate the Mode of action of Entomopathogenic Viruses?

■■■■

Section - 2
Practicals

Practical – 1

1. Objective

To know about Equipment used and Safety measures observed in Insect pathology laboratory.

2. Laboratory Equipment

Analytical Balances: An instrument used for accurate measurement/ determination of weight of a sample in various operations.

Care and maintenance of analytical balance

Never exceed the stated maximum load of the balance.

The balance must be kept clean.

Objects to be weighed should never be handled with the fingers.

Objects to weigh should be allowed to attain the temperature of balance.

Heating Apparatus: Ovens are the boxes designed to maintain a constant internal temperature; they are electrically heated and thermostatically controlled for drying the glass wares and other solids with very high melting points. Ovens are designed to maintain the differential internal temperature ranging from room temperature to about 250–300°C ± 1–2°C (Fig. 1.1).

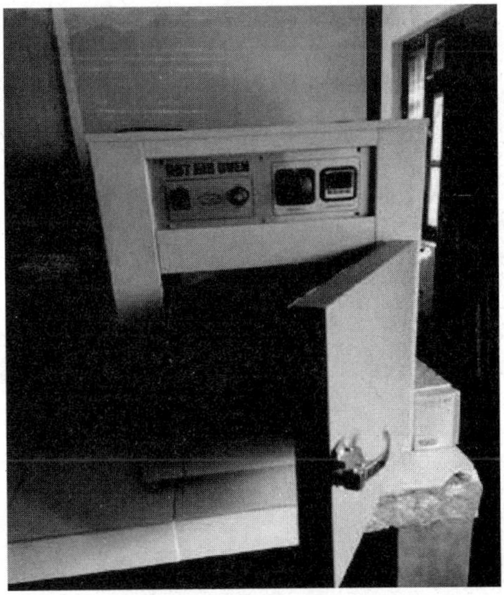

Fig. 1.1. Hot air oven

Incubators: They are the boxes designed to maintain a constant internal temperature by the use of thermostat; and are used for rearing insects and prematurely born infants. An incubator comprises a transparent chamber and the equipment that regulates its temperature, humidity and ventilation and; is used in biological experiments to allow growth of organisms in optimal conditions of temperature and humidity (e.g. BOD) (Fig. 1.2).

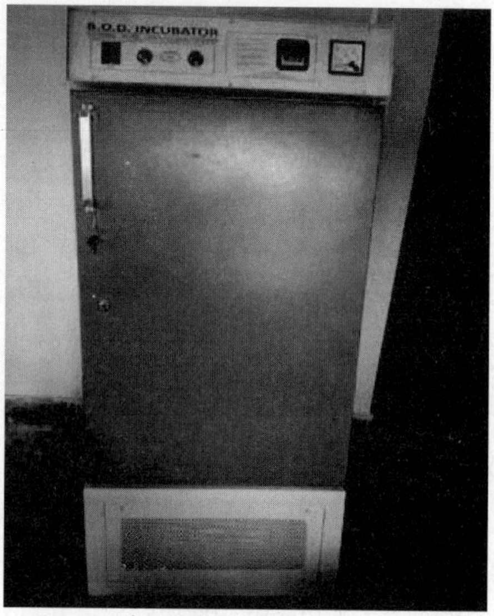

Fig. 1.2. Biological Oxygen Demand

Distillation Plant: It is an instrument used to purify the water through distillation. The working mechanism of the distillation plant allows the water to vapourized and condensed to form the distilled water (Fig. 1.3).

Fig. 1.3. Distillation Plant

Autoclave: A thick walled vessel with a tightly fitting lid, in which substances may be heated under pressure. It is used for sterilizing the media (Fig. 1.4).

Fig. 1.4. Autoclave

Crucibles and Beaker Tongs: Apparatus such as crucibles evaporating basins and beakers which have been heated need to be handled with suitable tongs.

Berlese Funnel: An apparatus comprising of funnel, lamp and collecting jar. It's an apparatus for collecting and extracting small invertebrate animals from grains, soil and decomposing litter.

Desiccators

A desiccators is a covered glass containers designed for the storage of objects in a dry atmospheres. It is charged with some drying agent such as silica gel, anhydrous calcium chloride. The exhausted silica gel (spent material) can be regenerated by heating in an electric oven at 150–180°C

Magnifying Instruments

i. Hand lens
ii. Pocket lens
iii. Microscopes: Microscope is an important instrument, which enlarge the image of an object with the help of lenses, so as to facilitate the morphological, anatomical and other characteristics of an object. It also facilitates in measuring sizes of smaller objects.

iv. Dissecting microscope, binocular microscope and electric microscope are common types of microscopes being used in laboratory. The smaller objects are measured by ocular and stage micrometres.

Care and maintenance of microscope

i. Carry the microscope in an upright position, by putting one hand on the handle and supporting the base with other hand.
ii. Keep microscope free from dirt and dust
iii. Always use clean glass slides with specimen on it.
iv. Reflect the mirror towards the light source
v. Always start focusing with low power to bring the object in the objective field. Use higher magnification gradually.
vi. Clean the microscope after finishing your work.
vii. Store the microscope in a cabinet with a cover over it

Glassware: In order to avoid the introduction of impurities during analysis, apparatus of resistance glass should be employed. For most purposes Pyrex glass (a borosilicate glass) is to be preferred. As a rule, glassware should not be heated with a naked flame, a wire gauze should be inter posed between the flame and glass vessel.

Mostly, used laboratory glass wares are as follows:

Beakers	Pipette	Glass rod
Measuring cylinder	Conical flasks	Glass jars
Desiccator	Petri dishes	Droppers
Specimen Vials	Reagent bottles	Wash bottles
Funnel	Separator funnel	Cover slips
Test tubes	Glass slides	

Chemicals : 1. Chemicals used for insect killing

Ethyl Acetate	Ammonia Water	Chloroform
Carbon Tetra Chloride	Potassium Cyanide	Calcium Cyanide
Diethyl Ether	Sodium Cyanide	Xylene + Ethyl Alcohol

Ethyl Acetate is the Active Ingredient commonly used to charge collecting jars.

Chemicals used as Preservatives

- Ethyl Alcohol
- Formaldehyde
- Glacial Acetic Acid

Chemicals used as Disinfectants

- Formaldehyde • Hypochlorite Acid • Benzene- Used for cleaning

Chemicals for Slide Preparation

- Xylene- xylene bath will clear the specimen by removing fats and oils
- Glycerine - for sealing the slides
- Potassium Hydroxide –remove the soft tissues
- Canada Balsom- fixing the specimen (mounting cement)

Chemicals for Pesticide Residue Extraction

Acetone, Acetonitrile, Hexane, Petroleum ether, Methyl chloride (Pesticide will be dissolved on Organic Solvents, particular type of Pesticides may require particular type of chemical)

Sodium Chloride – Help to separate aqueous layer and solvent layer distinctly

Sodium Sulphate – Moisture absorption from solvents if any

Florosil – Absorb pigments used on purification process

Acetone – Washing (Purification & Concentration)

Teepol – Teepol is a relatively mild and inexpensive detergent which may be used for cleaning glassware

3. Safety in the Laboratory

Each laboratory is unique in its operation and carries its own risk. It is the obligation of users to observe safety rules and regulations of each laboratory. Below are general safety guidelines applicable to most laboratories.

- No Smoking
- No Eating
- No Drinking
- Good Washing Facilities
- Good Disposal Facilities

General Precautions

a. Users are required to observe the necessary legislative ordinances when life specimens are involved in the experiment. Regulations should be followed in handling materials of potential biological and chemical hazards.

b. Personal protection equipment should be available and used in the laboratory whenever necessary.

c. Glassware or containers that had been used for laboratory operations should never be used for preparing or keeping food or drinks. Food storage is not permitted in laboratory freezers, refrigerators, ice chests, or cold rooms.

d. Follow disposal procedures for chemical wastes. Experimental apparatus may require traps or scrubbing devices to prevent the escape of toxic substances into the laboratory and the environment.

e. Make sure that all containers of hazardous substances (chemicals, biological agents or radioactive substances) are correctly and clearly labelled.

f. Always wash your hands and forearms with soap and water before leaving the laboratory area. Washing with solvents can remove the natural protective oil layer from the skin and can cause irritation and inflammation.

Precautions when dealing with chemicals

Always read the label on the container.

a. Orange background with black cross: harmful.

b. Orange background with black skull and Crossbones: toxic.

c. Orange background with black fire: flammable.

d. Orange background with black Bar or hand being eaten away: corrosive.

e. ALWAYS keep a list of chemicals and the dangers associated with them.

f. NEVER store chemicals in anything other than their original container, with the correct label.

g. ALWAYS keep flammable and toxic chemicals in a securely locked cupboard.

Precautions when dealing with Biological Hazards

a. The entomopathogenic fungi used for locust control are very safe, BUT they may cause allergic reactions if they are inhaled.

b. Some common contaminants *e.g. Aspergillus* sp. Are harmful.

c. Technicians must learn to recognise and dispose of contaminants safely.

d. ALWAYS keep benches, equipment and protective clothing clean.

First Aid

a. ALWAYS keep a first aid box in a secure place, in full view and check it regularly.

b. Write all injuries in an accident book which should be kept next to the first Aid box.

c. It is ESSENTIAL that at least one person in a laboratory is trained in first aid procedures.

d. KEEP AN ANTIDOTE FOR ANY DANGEROUS CHEMICALS

4. Reference

Lamsal M.P. and Paudel P. (2015). Manual on Laboratory Handling Techniques for Plant Quarantine Services. *National Plant Quarantine Program*, Lalitpur, Nepal. pp.54

■■■■

Practical – 2

1. Objective

Identification and Symptoms of Infection – Bacterial insect pathogens.

2. Description

Bacteria are microscopic single-celled organisms that thrive in diverse environments. They can live within soil, in the ocean and inside the human gut. Bacteria are prokaryotes. The entire organism consists of a single cell with a simple internal structure. Unlike eukaryotic DNA, which is neatly packed into a cellular compartment called the nucleus, bacterial DNA floats free, in a twisted thread-like mass called the *nucleoid*.

Bacterial cells also contain separate, circular pieces of DNA called *plasmids*. Bacteria lack membrane-bound organelles, specialized cellular structures that are designed to execute a range of cellular functions from energy production to the transport of proteins. However, both bacterial and eukaryotic cells contain ribosomes. These spherical units are where proteins are assembled from individual amino acids, using the information encoded in a strand of messenger RNA.

On the outside, bacterial cells are generally surrounded by two protective coverings: an outer cell wall and an inner cell membrane. However, certain bacteria, like the mycoplasmas do not have a cell wall at all. Some bacteria may even have a third, outermost, protective layer called the *capsule*.

- **Groups:** Spore forming and non-spore forming bacteria.
- **Reproduction:** Bacteria grow independently in the insect haemocoel.
- **Infection:** Bacteria must be ingested. They rupture the gut wall, then invade the haemocoel.
- **Survival:** Some bacteria form spores which are moderately resistant.
- **Host range:** Lepidoptera, Coleoptera and Diptera.
- **Culture:** They can easily be grown in petri dishes or in liquid culture but specific conditions are needed for spore information in artificial media.
- **Biocontrol agents:** The main bacterial control agents of insects are species from the genus ***Bacillus***. Other groups exist, but typically only cause disease in weak, injured or stressed insects.

I. *Bacillus* sp.

- **Description:** These are aerobic, unicellular, usually bacilliform, spore forming bacteria.
- **Infection**: Infection occurs after ingestion of bacterial cells or spores. They mainly affect phytophagous larval stages. Bacteria invade and reproduce in the haemocoel, inducing lethal septicaemia.
- **Culture:** Easily produced by fermentation.
- **Symptoms:** The bacterial toxin attacks the gut lining and causes muscle paralysis in the alimentary tract and the mouth parts. Feeding stops. Insects may regurgitate and have diarrhoea.
- **Death of host:** Between 30 minutes and 24 hours. Death may be caused by starvation.
- **Susceptibility:** Susceptibility depends on the gut of the host being alkaline.
- **Locusts:** No strains of *Bacillus* are known to affect locusts, although there is a report of an NPV infection.
- **Biocontrol agents:** There are several species of Bacillus which are used in biological control. Important ones are discussed herein.

i. *Bacillus thuringiensis:* During sporulation the bacteria produce a large proteinaceous crystal which is bi-pyramidal in shape and also a thick walled endospore. The crystal is an endotoxin which dissolves inside the host in alkaline gut fluids and releases toxic polypeptides. Mainly Lepidoptera are attacked, but strains specific to mosquitoes and Coleoptera are also known. It has been developed as a commercial product, widely available throughout the world.

ii. *Bacillus sphaericus:* Certain strains produce a proteinaceous toxin which poisons mosquito larvae. Host death may also be induced by the spores alone.

iii. *Bacillus popilliae*: Produces no toxins in the infection cycle. Japanese beetle (*Popillia japonica*) larvae ingest spores in the soil which germinate and produce vegetative cells which fill the gut in three to five days. Some cells penetrate the gut wall, then grow and sporulate in the haemolymph. 14–21 days after initial infection the insect body is swollen and creamy white (milky disease). After death, the spores are released into the soil and establish persistent infection sites.

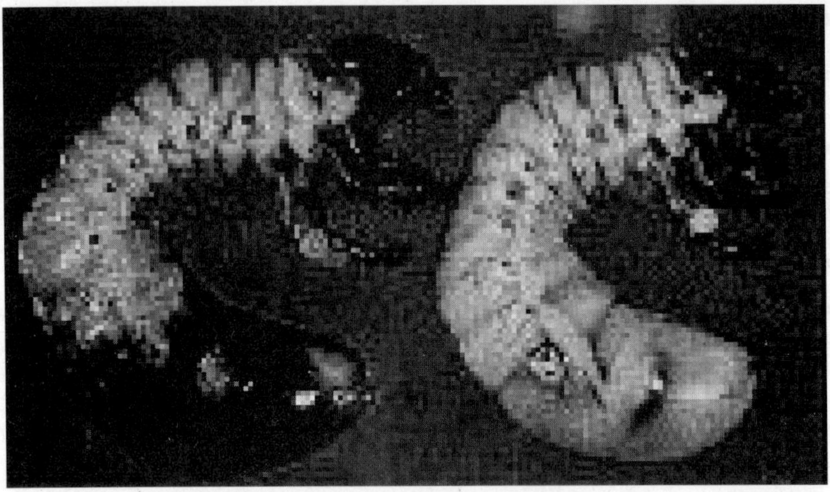

Fig. 2.1. Healthy beetle larva (left) larva infected with *Bacillus popilliae* (right).

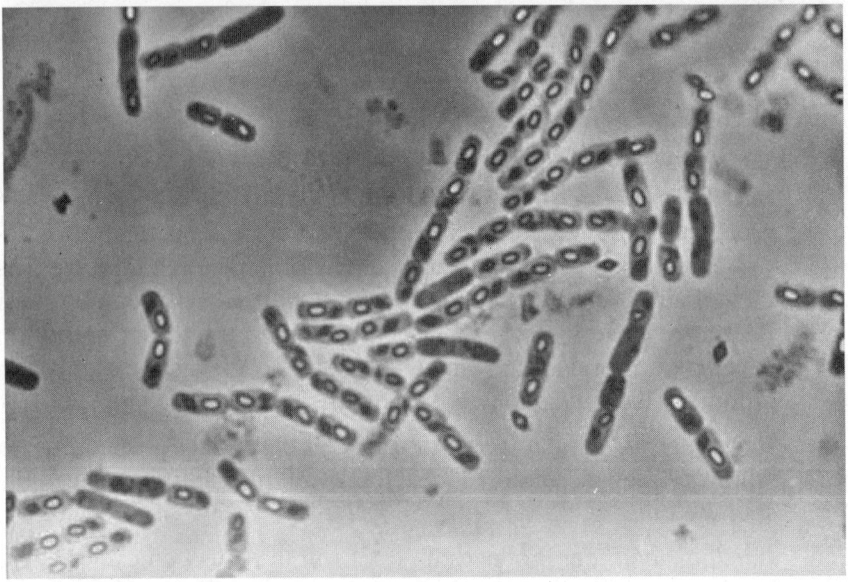

Fig. 2.2. *Bacillus Thuringiensis* microscopic view. P.C. Indiamart.

Identification (external symptoms)

To determine the cause of death of an insect, first look at the insect or external symptoms may tell you which kind of pathogen is responsible for death. In case of bacterial infection:

i. Larvae remain normal colour.

ii. After death darken to brownish-black.

iii. Often flabby (Not strong or robust enough to hold the shape).

iv. Generally does not leak any watery fluid out of body.

Further Identification (Elaborated examination)

i. Before the insect decomposes, examine a drop of haemolymph under a phase contrast microscope.

ii. If the bacteria are saprophytic you will see bacterial cells of different shapes and sizes.

iii. If there are concentrations of bacilli-form cells this indicates the possible presence of an entomopathogenic bacterium.

Note: Insect gut normally contains saprophytic bacteria. You must take care not to mistake these for disease-causing bacteria.

iv. ***Bacillus thuringiensis*:** Good growth on nutrient agar, catalase positive, crystal parasporal body present.

v. ***Bacillus popilliae*:** Little or no growth on nutrient agar, catalase negative, crystal parasporal body usually present.

vi. ***Bacillus sphaericus*:** Spores are almost spherical. No crystalline parasporal body.

We cannot corroborate the inferences, until we get assured by isolating and culturing the bacteria associated with diseased insect to determine whether bacteria are involved in insect disease or not. However, that part has been discussed in chapter 9 of this book.

3. References

https://simple.wikipedia.org/wiki/Bacteria

https://www.indiamart.com/proddetail/bacillus-thuringiensis-8656353591.html

https://www.livescience.com/51641-bacteria.html

Practical – 3

1. Objective

Identification and Symptoms of Infection – Fungal insect pathogens.

2. Description

Fungi (singular: fungus) are a kingdom of usually multicellular eukaryotic organisms (with a well-defined nucleus and organelles, characterised by chitinised cells) that are heterotrophs and have important roles in nutrient cycling in an ecosystem. The cells are formed into filaments or hyphae together forming a mycelium. Fungi reproduce both sexually and asexually, and they also have symbiotic associations with plants and bacteria. However, they are also responsible for some diseases in plants and animals including insects (Arthropods). They are simple, aerobic organisms (such as mildews, molds, mushrooms, smuts, toadstools, and yeast) which –

1. Unlike bacteria can grow in low moisture and low pH environments, and have their genetic material bound in a membrane, and
2. Unlike plants do not have roots or leaves, do not contain chlorophyll, and do not produce their own food, but obtain nourishment from dead organic matter. Out of about 70,000 species of fungi some are edible but most are poisonous to humans.
3. Their principal commercial uses include manufacture of food and alcoholic beverages, and antibiotics such as penicillin and in recent year's involvement in IPM as biocontrol agents against insect pests.

Groups: 1. Entomophthorales, 2. Deuteromycetes

Entomopathogenic fungi are found in several subdivisions. The fungi in the Entomophthorales (*e.g. Entomophaga grylli)* have complex life cycles involving a sexual stage and resting spores, whereas Deuteromycete fungi (*e.g. Metarhizium* and *Beauveria*) have a simple life cycle with no known sexual stage.

- **Reproduction:** Fungi reproduce by spores, which are formed sexually or asexually.
- **Infection:** Spores germinate on the host cuticle and penetrate using enzymes and mechanical pressure. Inside the haemocoel the fungus multiplies rapidly by budding or hyphal fission. The resulting yeast-like cells (blastospores) spread through the body.
- **Death of host:** By extensive mycelial colonisation causing asphyxiation or starvation, or by toxins released in the yeast phase. The

cadaver desiccates as the hyphae use host nutrients and water to develop.

- **Survival:** Hyphae break through the cuticle after death. Spores may be liberated passively or actively to continue the infection cycle.

1. Entomophthorales

- **Reproduction:** Entomophthoralean fungi have complex life cycles involving non-sexual conidia, and sexual resting spores.
- **Culture:** Entomophthorales fungi must be cultured in complex media and some cannot be cultured in artificial media. Some produce mycelium, but do not sporulate.

Biocontrol agents

i. *Entomophaga praxibulli,* has been introduced to the USA for control of Grasshoppers.

Fig. 3.1. Grasshopper killed by a complex of Entomophaga species

Fig. 3.2. Sporulating grasshopper cadavers killed by *Entomophaga grylli*

ii. *Entomophthora plutellae,* used against *Plutellai* in Thailand.

iii. *Zoophthora radicans,* has been used as a classical biocontrol agent against aphids in Australia.

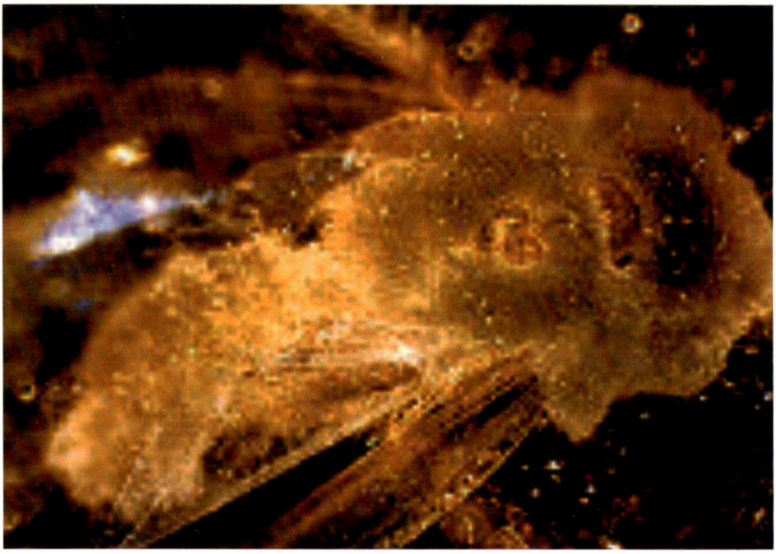

Fig. 3.3. Leafhopper infected by *Zoophthora radicans*

Fig. 3.4. Locusts attacked by the fungus *Metarhizium*

2. Deuteromycetes

- **Reproduction:** The Deuteromycete fungi have no known sexual stage. Although they do have mechanisms for the exchange of genetic material, most reproduction is asexual, by the production of asexual spores called *conidia*.
- **Culture:** Many deuteromycetes are easily cultured in simple artificial media.

Fig. 3.5. An aphid killed by an entomopathogenic fungus (*Verticillium lecanii*)

Biocontrol agents

i. ***Metarhizium*** : *M. anisopliae*, and the sub-group *M. flavoviride* have been produced on a large scale on rice substrate in several tropical countries.

ii. ***Beauveria*** : *B. bassiana* and *B. brogniarti* have been produced on a large scale in China, Europe and America.

iii. ***Verticillium lecanii,*** has been developed as a biopesticide for use in glasshouses in Europe.

Fig. 3.6. Fungal pathogen *Beauveria bassiana* on cadaver of western flower thrips.

Identification (external symptoms)

To determine the cause of death of an insect, first look at the insect or external symptoms, may tell you which kind of pathogen is responsible for death. In case of fungal infection:

i. All growth stages of an insect can be affected.

ii. After death the cadaver is desiccated, never flaccid. (Unlike bacteria)

iii. In soft-bodied insects, the body is covered with fungal mycelium.

iv. In hard bodied insects bands of hyphae grow between the integuments and mycelial strands grow from the orifices fixing the insect to the substrate.

v. Spores may form on the outside of the insect, but if the environment is very dry, the fungus may not show on the outside of the insect.

Further Identification (Elaborated examination)

i. Entomopathogenic fungi sporulate on the outside of the host insect under moist conditions and on the inside of the host when the environment is too dry.

ii. Deuteromycetes fungus: indicated by the presence of masses of powdery spores.

   Beauveria sp.: white spores.

   Metarhizium sp.: green spores. Prominent hyphal structures, often brightly coloured, ascomycete (sexual stage) or an asexual genera *e.g., Hirsutella.*

iii. Entomophthorales fungus: Stout hyaline hyphae surrounded by halos of white spore deposits.

Note: You must also use the microscope to confirm identification.

We cannot corroborate the inferences, until we get assured by isolating and culturing the bacteria associated with diseased Insect to determine whether bacteria are involved in insect disease or not. However, that part has been discussed in chapter 9 of this book.

3. References

https://biologydictionary.net/fungi/

http://www.businessdictionary.com/definition/fungi.html

https://biocontrol.entomology.cornell.edu/pathogens/entomophaga.php

https://www.mindenpictures.com/search/preview/an-aphid-killed-by-a-entomopathogenic-fungus-verticillium-lecanii/0_80125691.html

http://biocontrol.ucr.edu/wft.html

Montserrat, M., C. Castañé and S. Santamaria (1998). *Neozygites parvispora* (Zygomycotina: Entomophthorales) causing an epizootic in *Frankliniella occidentalis* (Thysanoptera: Thripidae) on cucumber in Spain. Journal of Invertebrate Pathology 71: 165-168.

■■■■

Practical – 4

1. Objective

Identification and Symptoms of Infection – viral insect pathogens.

2. Description

Viruses are sub-microscopic, intracellular, obligate pathogens. They can neither move nor metabolise. They consist of a template nucleic acid with or without a protein coat, which is also called an inclusion or occlusion body.

3. Reproduction

Viruses multiply by independent synthesis of their component parts. These parts assemble to produce progeny virus within the host cell.

4. Infection

Viruses normally enter the host through the mouth. The protein coat of the virus dissolves in the gut, releasing the virus particles (virions). The virions invade, then multiply in the cells of the gut wall. Replication on a massive scale then takes place in the fat body, haemocytes and hypodermis.

5. Death of host

Death usually occurs in three to ten days.

6. Survival

After death the body ruptures and releases millions of occlusion bodies. Occlusion bodies protect the virus which can persist for years in the right conditions.

7. Host range

The host range is narrow, but viruses are known from a wide range of hosts.

8. Culture

Viruses can only be cultured in the live insect host or in tissue culture.

9. Groups

Baculovirus, entomopox virus, picornavirus, cytoplasmic polyhedrosis virus.

10. Biocontrol agents

At present, the main group of viruses used as biological control agents are the baculoviruses are discussed herein.

BACULOVIRUSES

- **Description:** Baculoviridae are the safest insect viruses to use as pathogens, since no similar viruses are known to infect vertebrates or plants. They have double-stranded DNA and are protected by a protein coat which improves their persistence.
- **Infection:** Infection occurs after susceptible insect larvae eat food contaminated with virus. The virus then attacks the haemolymph, fatty tissue and mid gut. The insect becomes paralysed.
- **Virulence:** Highly virulent; the presence of very few particles can initiate infections and hosts die within 3–10 days.
- **Susceptibility:** The gut of the host insect must be alkaline so that the occlusion body can dissolve.
- **Locusts:** No Baculovirus has been recorded from locusts with the exception of an unconfirmed report of cross infection from a *Spodoptera* sp. (Lepidoptera) host.
- **Groups:** Three groups of baculoviruses are described below:

  1. Nuclear polyhedrosis viruses,
  2. Granulosis viruses and
  3. Group C baculoviruses.

1. Nuclear polyhedrosis viruses (NPV)

1. **Description**: About 280 species are known. Rounded cubic or hexagonal polyhedra. 0.5–1.5 microns (μm) in size. Singly or multiply enveloped (see Figure 1).
2. **Infection:** Infection occurs in the adipose tissue of the hypodermis, in the tracheae and the middle intestine.
3. **Host:** Approximately 120 species of Lepidoptera and Hymenoptera (particularly saw-flies). Each virus is highly specific to its host.
4. **Survival:** Nuclear polyhedrosis viruses form particles inside a crystalline protein structure (occlusion body). This allows the virus to survive outside the host for years out of sunlight.
5. **Biocontrol agents:** The following NPVs have all been produced on a commercial or semi-commercial scale: *Autographa californica* NPV, *Lymantria dispar* NPV, *Malacasoma disstria* NPV, *Mamestra*

brassicae NPV, *Neodiprion sertifer* NPV, *Spodoptera* NPV and *Heliothis* NPV.

2. Granulosis viruses (GV)

- **Description:** About 65 species Oval or ovoid granules.
- **Infection:** Granulosis viruses attack the adipose tissue.
- **Host:** Lepidoptera larvae.
- **Survival:** Granulosis viruses form particles inside a crystalline protein structure (occlusion body). This allows the virus to survive outside the host. Can survive for years out of sunlight.
- **Biocontrol agents:** Include *Cydia pomonella* GV (codling moth), *Phthorimaea operculella* GV (potato tuber moth).

3. Group C Baculoviruses

- **Description:** Double stranded DNA. Viruses with non-included virions. Only visible using an electron microscope. 22–30 nm in size. These viruses are unusual since they have no protective protein coat to help them to survive.
- **Infection:** These viruses attack the haemolymph, fat body, midgut. Insects become paralysed.
- **Host:** These viruses are restricted to Arthropoda. Larvae and adults of Coleoptera, Hymenoptera and mites.
- **Biocontrol agents:** *Baculovirus oryctes*: Used for the control of rhinoceros beetles, *Oryctes* sp. This virus is excreted from the living diseased insect as virions. These are passed on to other adults during mating. Some spread occurs from contamination of adult breeding and larval feeding sites, but the virus does not survive long in the environment.

4. Entomopox Viruses

- **Description:** Entomopox viruses have inclusion bodies (i.e. they are occluded viruses) which are important for identification. Spherical or ovoid particles. 5–20 μm in size.
- **Infection:** These viruses must be ingested by the host. They then attack the fat body.
- **Virulence:** They kill hosts more slowly than baculoviruses.
- **Host:** Lepidoptera larvae, Diptera, Coleoptera, Orthoptera.
- **Survival:** Little is known.

- **Locusts:** Entomopox viruses have been recorded from locusts and grasshoppers.

Identification (external symptoms)

To determine the cause of death of an insect, first look at the insect or external symptoms may tell you which kind of pathogen is responsible for death. In case of viral infection:

i. Infection by viruses mainly occurs in the larval stage.

ii. Larvae become pale and flaccid. Dark in colour after death.

iii Baculovirus infections:

   a. Body contents become liquid.

   b. Larvae may hang by their prolegs.

   c. May ooze white fluid.

   d. Infected larvae are sometimes smaller than healthy larvae.

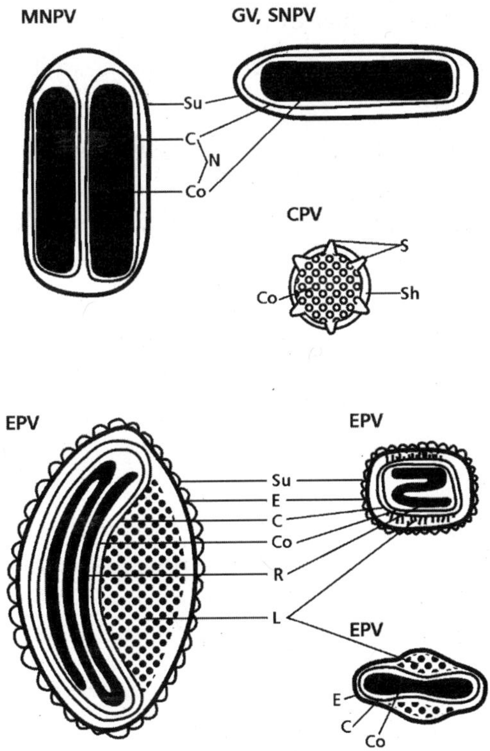

Fig. 4.1. ***C*** capsid; ***Co*** core; ***CPV*** cytoplasmic polyhedrosis virus; ***E*** envelope; ***EPV*** entomopoxviruses; ***GV*** granulosis virus; ***L*** lateral body; ***MNPV*** multiple nucleocapsides per envelope nuclear polyhedrosis virus; ***N*** nucleocapsid; ***R*** colled rodlike structure; ***S*** splkes; ***Sh*** shell; ***SNPV*** single nucleocapsid per envelope nuclear polyhedrosis virus; ***Su*** surface units

Fig. 4.2. Dead Caterpillar Killed By NPV. P.C. Kathy Keatley Garvey

Further Identification (Elaborated examination)

Occluded insect viruses are the most commonly found viruses.

a. Use a bright field or phase contrast microscope.

b. You should see the characteristic shining white (mono-refrainment) inclusion bodies.

c. Use stains such as Giemsa and Feulgen-Schiff to confirm presence of inclusion bodies.

d. Further identification of non-occluded viruses must be done using electron microscopy and serology techniques.

5. Reference

http://ucanr.edu/blogs/bugsquad//blogfiles/25241_original.jpg

Practical – 5

1. Objective

To understand the concept of Pathogen Load.

2. Definition

Term "Pathogen load" is a generalised indicator that quantify number of infective or non-infective pathogen in the host body after certain standard gestation period.

3. Introduction

It seems obvious that the number of pathogens should be important in the pathogenesis of an infectious disease (Medzhitov *et al.,* 2012). The relationship between the pathogen load and severity is one of the most fundamental questions, and yet, strangely, one of the most difficult to answer. One reason for this is that it is often rather difficult to determine the total pathogen load in an infected host, particularly in an infected insect. Pathogens can be distributed, non-uniformly, throughout multiple different cell, tissue, or organ, compartments of the body, many of which are difficult to sample.

For this reason, we are often constrained by measuring pathogen load in the samples that are readily accessible such as haemolymph, internal body organs like mid gut, and external tissues like cuticle or tergum/notum *etc*., and we must assume that these are representative of total pathogen load. The success of this approach is borne out by the usefulness of these measurements to guide management of insect colonies with some infections, but it is also well recognised that these methods are imperfect (Ruell *et al.,* 2007). Variations in both pathogen and host can alter distribution of the pathogen, and the likelihood that a given pathogen burden will cause disease (Medzhitov *et al.,* 2012; Schneider and Ayres, 2008).

Imperfect estimation of pathogen load becomes a particularly important problem when trying to understand host responses to infection, and their role in pathogenesis. In order to interpret whether a host response is insufficient, appropriate, or excessive, it needs to be assessed in relation to the pathogen load that triggered it.

4. References

Medzhitov, R., Schneider, D.S. and Soares, M.P. (2012) Disease tolerance as a defence strategy. *Science* 335: 936–941.

Ruell, J., Barnes, C., Mutton, K., Foulkes, B., Chang, J., Cavet, J., Guiver, M., Menasce, L., Dougal, M. and Chopra, R. (2007) Active CMV disease does not always correlate with viral load detection. *Bone Marrow Transplant* 40: 55–61.

Schneider, D.S. and Ayres, J.S. (2008) Two ways to survive infection: what resistance and tolerance can teach us about treating infectious diseases. *Nat Rev Immunol* 8: 889–895.

■■■■

Practical – 6

1. Objective

To know the different techniques for extraction of fungal pathogens from live or dead insect/organism.

2. General Principle

Isolation of the fungal pathogens from diseased insect is made by surface sterilizing the diseased area with surface sterilizing agents, removing a small portion of the infected tissue (Tergum (Notum), Sternum, Pleuron etc.) with a sterile scalpel, and plating it in a plate containing a nutrient medium. The most common method, for isolating fungal pathogens from infected larvae as well as other stages of insect involves cutting several small sections (5–10 mm) from the infected parts to contain both diseased and healthy looking tissue. These are placed in surface sterilizing agents' solutions for about 15–30 seconds. The sections are taken out aseptically and blotted dry on clean, sterile paper towels or washed in three changes of sterile water and are finally placed on the nutrient medium, usually three to five per dish. The pathogen will grow from the sections and the colonies of the pathogen are sub cultured aseptically for further study.

3. Materials Required

Infected specimens (Cadavers of arthropods should be collected from pesticide free areas), sterile petri-dishes, PDA slants, sodium hypochlorite solution (1%), sterile water, razor blade, forceps, inoculation needle, burner/spirit lamp, spirit, incubator, PDA medium.

4. Isolation and Identification: Protocols

1. Select infected host tissue from the affected insect's body parts.
2. Cut into small pieces (2–5 mm) containing both the diseased and healthy tissue and keep in sterile petri dishes.
3. Dip the pieces into 1% sodium hypochlorite solution for about one minute.
4. Transfer the pieces to petri dishes containing sterile distilled water and wash thoroughly in two changes of sterile water to free them from the chemicals if any.
5. Wash hands with rectified spirit and wipe the table top of inoculation chamber with rectified spirit.

6. Lit the burner
7. Hold the flask containing sterile Luke warm PDA in the right hand and remove plug near the flame. Lift the lid of petri dish gently with left hand and pour about 20 ml of medium. Close the mouth of the flask with plug near the flame.
8. After solidification of the medium, place four sterilized pieces at different distance in a single PDA plate.
9. Incubate the petri dishes in an inverted position at 25ºC and examine for 3–5 days.

5. Observations

Observe the incubated plates from the second day onwards for the growth of the fungus. Aseptically transfer the bits of mycelia from the margin of the colonies on fresh PDA slants for further study, sporulation and identification. Mycelia growth on the medium from the infected tissues indicates that the disease may be due to a fungus.

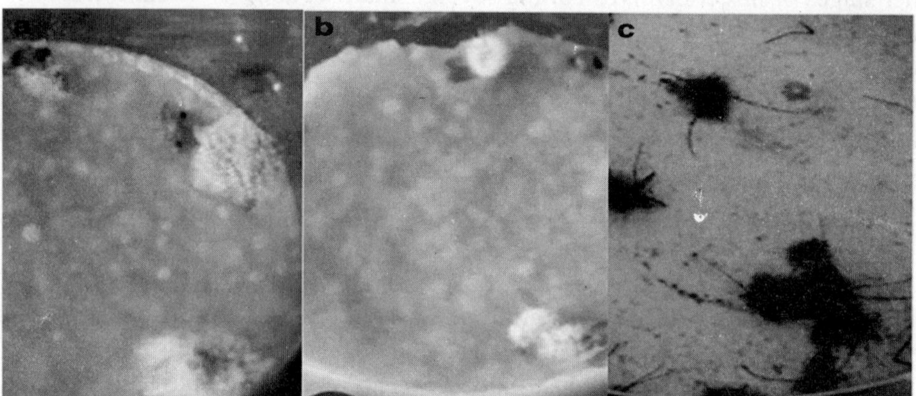

Fig. 6.1 Photos of dead insects after microinjection or spraying with fungal entomopathogenic spores. (a) Drosophila cadaver. (c) Aedes mosquitoes after spraying with the same fungus. The dead insects are completely covered by fungal growth indicating that the cause of death is the development of the spores within the Insect. (b) Drosophila cadavers after *B. bassiana* spores micro-injection are shown for comparison. [P.C. Jaber *et al*., 2016]

6. Reference

Jaber, S., Mercier, A., Knio, K., Brun, S. and Kambris, Z. (2016). Isolation of fungi from dead arthropods and identification of a new mosquito natural pathogen. *Parasites & Vectors*, *9*(1): 491.

Practical – 7

1. Objective

To know the different techniques for extraction of bacterial pathogens from live or dead insect/organism.

2. General Principle

Isolation and identification of bacteria associated with diseased insect is important to determine whether bacteria are involved in insect disease. The method normally used to isolate Entomopathogenic bacteria differs from that used for fungi. A suspension of bacterium is prepared from the infected material and loopfuls of this are streaked onto nutrient agar plates. The aim is to produce single colony that can be sub-cultured pure.

3. Materials required

Disease specimens (Larvae and adults of the common Arthropods *viz.* grasshoppers, crickets, cockroaches etc. from pesticide free areas), growing media (petri-plates) surface sterilizing agents (1% sodium hypochlorite), sterile razor blade, glass rod, sterile water, sterile test tubes and petri-dishes, sterile pipettes (1 ml), inoculation loop.

4. Isolation and Identification: Protocols

1. Living and dead larvae and adults of field-collected insects should be surface sterilised with 70 per cent alcohol and homogenised in a homogeniser containing 5 ml of sterilised phosphate buffer solution (PBS, pH 7.4).
2. Bacterial isolates are prepared in one of two ways. Isolation of bacteria should be done from haemolymph or in other case, the insect body must be triturated.
3. The suspensions which were obtained from larvae and adults bodies were filtered twice through two layers of cheese cloth to remove debris. (Poinar, 1978).
4. Dilute the suspension to 10^{-8} (Christine & Ted, 1992).
5. Heat the suspension at 80°C for 10 min in a water bath to eliminate the nonspore-forming bacteria (Ohba & Aizawa 1986; Lee *et al.* 1992; Thiery & Frachon 1997).

6. Spread 100 µl of the heat-treated samples on nutrient agar plates and incubate at 28°C for 48–96h (Lee *et al.,* 1995).
7. Alternately, apply the suspension directly to nutrient agar plates.
8. Then examine and select bacterial colonies based on colour and morphology.
9. Prepare pure cultures of bacterial colonies, and identify these cultures by various tests, such as use of organic compounds, spore formation, NaCl tolerance, optimum pH, optimum temperature, catalase test, oxidase test, and gelatine hydrolysis.
10. The identification of isolated bacteria should be done according to "bergey's manual of Systematic Bacteriology 1 and 2" (Krieg & Holt 1986; Sneath *et al.*, 1986).

Serial dilution method: The following figure 7.1 is the technique for serial dilution method.

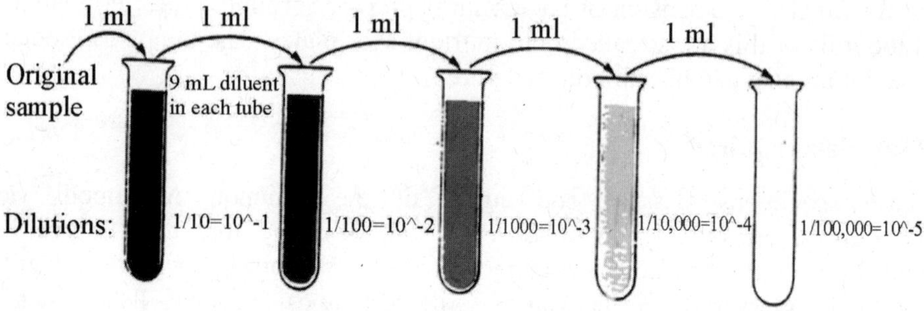

Fig. 7.1 Serial Dilution Method.

Plate count method: First pour the bacterial cell suspension in sterilized Petri plate containing medium ranging from 0.1 ml to 1 ml depending upon population of plant pathogenic bacteria. Take sterilized medium in 40–50ºC, 10–15 ml per sterilized plate media is poured to each plate. Incubate the material in BOD (Biological Oxygen Demand) incubator 25±2ºC.Observe the development of colony and pick up the desired colony, then purify the culture using subsequent transfer and streak method. Maintain the pure strand and keep them in refrigerator.

5. References

Christine, L.C. and Ted, R.J. (1992). Laboratory experiments in microbiology. California, Benjamin/Cummings.

Krieg, N.R. and Holt, J.G. (1986). Gram-negative aerobic rods and cocci. In: Palleroni NJ ed. Bergey's manual of systematic bacteriology. Baltimore, Williams and Wilkins. Pp. 140–218

Lamsal, M.P. and Paudel, P. (2015). Manual on Laboratory Handling Techniques for Plant Quarantine Services. *National Plant Quarantine Program*, Lalitpur, Nepal. pp. 54.

Lee, H.H., Joo, D.G., Kang, S.C. and Lim, H.G. (1992). Characterization of *B. thuringiensis* — seven isolates from soil (I). *Korean Journal of Applied Microbiology and Biotechnology* 20: 377–388.

Lee, H.H., Jung, J.D., Yoon, M.S., Lee, K.K., Lecadet, M.M., Charles, J.F., Dumanoir, V.C., Franchon, E. and Shim, J.C. (1995). Distribution of *Bacillus thuringiensis* in Korea. *B. thuringiensis Biotechnology and Environmental Benefits* 1: 201–215.

Ohba, M. and Aizawa, K. (1986). Distribution of *Bacillus thuringiensis* in soil of Japan. *Journal of Invertebrate Pathology* 47: 277–286.

Poinar, G. (1978). *Identification of the groups of insect pathogens.* New York, Plenum Press.

Sezen, K., Demir, Ý. and Demirbağ, Z. (2007). Identification and pathogenicity of entomopathogenic bacteria from common cockchafer, *Melolontha melolontha* (Coleoptera: Scarabaeidae). *New Zealand journal of crop and horticultural science, 35*(1): 79–85.

Sneath, P.H.A., Mair, N.S., Sharpe, M.E. and Holt, J.G. (1986). Regular, nonsporing gram-positive rods. In: Kandier O, Weiss N ed. Bergey's manual of systematic bacteriology. Baltimore, Williams and Wilkins. Pp. 1208–1260.

Thiery, I. and Frachon, E. (1997). Identification, isolation, culture and preservation of entomopathogenic bacteria. In: Lacey AL ed. *Manual of techniques in insect pathology.* London, Academic Press. Pp. 55–73.

■■■■

Practical – 8

1. Objective

To know the different techniques for extraction and storage of Viral pathogens from live or dead insect/organism.

2. Introduction

Virus isolates can be maintained for long periods by sub-culturing on their respective hosts, but may become contaminated or attenuated during repeated sub-culturing. It is advisable, therefore, to store samples of the original isolates as soon as possible after they have been acquired. When a fresh isolate is obtained, its concentration should be built up by sub-culturing once or twice in a suitable host insect. Then it should be preserved by any one of the following methods.

a. As dried powder

- Place infected insects over anhydrous calcium chloride in an airtight, screw topped bottle for several months.
- Grind the dry insects to a powder with a pestle and mortar, and store in a tightly sealed capsule.

b. In liquid nitrogen

- Grinding infected insects in K_2HPO_4/Na_2SO_3 solution as if preparing inoculum for sap transmission.
- Place 0.5 ml aliquots of sap in sealed capsules and store under liquid nitrogen in a commercial storage flask.

c. By Lyophilisation

i. Take a small piece of infected insect and place in a glass ampoule suitable for the freeze-drying machine to use. Add a label to identify and date the isolate. Constrict the tube and evacuate the air. Seal the tube and store at room temperature.

ii. Alternatively, grind infected insect in a minimal volume of $K_2HPO_4/$ Na_2SO_3 solution. Take 10 ml of the filtered sap and add 0.7 gm D-glucose + 0.7gm peptone. Shake the mixture to dissolve the additives. Pipette 0.25 ml aliquots of the mixture in to the freeze drying ampoule. Add a small plug of non-adsorbent cotton wool, followed by a suitable paper label. Proceed the Lyophilisation as described above.

4. Procedure for dry preservation

First collect and place viral infected specimens in laboratory by pressing in newspaper and in open air to dry these samples. After a week, after complete drying, grind sample to a powder form with mortar and pestle.

Then place sample in airtight bottle with calcium chloride ($CaCl_2$).

The following is the purification method of insect virus

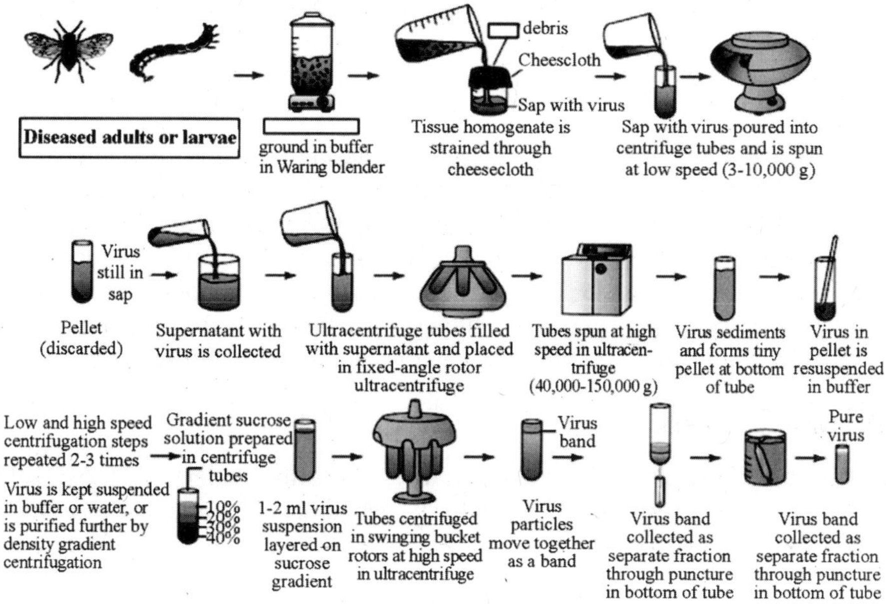

Note: The viral isolate so extracted and stored can be re-inoculated on suitable hosts for counterchecking their virulence and viability over a period or as and when felt necessary.

5. Reference

Lamsal M.P. and Paudel P. (2015). Manual on Laboratory Handling Techniques for Plant Quarantine Services. *National Plant Quarantine Program*, Lalitpur, Nepal. pp. 54.

Practical – 9

1. Objective

To know the different laboratory bioassay techniques for evaluation of different insecticides against insect pests.

2. Introduction

Laboratory bioassays now a days became increasingly crucial because of their predictive value, generating comparative toxicity data on many chemicals in relatively short times at relatively low expense. The laboratory investigations provide a better understanding of insect-insecticide or insect-plant-insecticide interactions. It is a simple, versatile, easy and sensitive technique for determining toxicity of wide range of chemicals, which greatly facilitates the determination of the LD50, LC50 or any other lethal concentration/dose (Burchfield *et al.,* 1952).

3. The Basic Principles

The basic principles of bioassay studies to evaluate the toxicity of insecticides with diverse mode of action to the same species under the same test conditions. Due to increasing uses of highly toxic insecticides to control destructive insect pests, it is important to analyse not only the technical and formulated materials but also minute quantities of their residues on/in plant and animal tissues (Sun, 1963). In an effective bioassay, the indicator species should be sufficiently sensitive to detect even small amounts of insecticides and should express the response with increasing concentrations (Chhillar *et al.,* 2007). The toxic interactions of an insecticide with a biological system are dose dependent. The toxicity of an insecticide to an organism is usually expressed in terms of LD50 (lethal dose). This value represents the dose per unit weight lethal to 50 per cent of the population of the organism (Finney, 1971). The LD50 is commonly expressed as milligram per kilogram (mg/kg). Bioassay methods, on the other hand, are highly sensitive, simple in operation, easily adapted to the assay of newer insecticides and capable of detecting toxic metabolites. The main purpose of these bioassays is to select newer insecticides and their most appropriate doses that affect insects, as well as to test pest resistance (Siqueira *et al.,* 2000) and the pesticide selectivity to natural enemies (Bacci *et al.,* 2009). Here the attempt was made to review the different laboratory bioassay methods available to evaluate the pesticide toxicity against insect pests.

4. Testing Protocols

Batches of insects are exposed to dosages of an insecticide. The insects to be tested should be free from variations due to age, stage, sex, condition of nutrition, etc. With precise experimental conditions, batches as small as 10 to 20 individuals may be used and it is desirable to replicate the batches 3 to 5 times. The toxicity of an insecticide is usually determined using a dose response relationship. The animals available for testing are divided at random into several groups, usually six. One of these is used as the control group to be treated with solvent only. The other groups are treated with the test compound with doses usually in a geometric progression (*i.e.*, 1, 2, 4, 6 or 1, 3, 9, 27 *etc.*). After the insects are exposed to different concentrations of an insecticide, they may or may not be kept in another container having food depending upon the method employed. The mortality count is taken after an interval of time already calibrated at different time intervals and when it becomes constant at particular period. The period should be fixed an exposure period. This can be done by exposing the batch of insects in control (where no insecticide is applied) exactly in the same manner as is done in exposure to insecticide. When there is natural mortality among the controls, the mortalities have to be corrected by Abhott's formula (Abbott, 1925) as follows

$$\text{Corrected \% mortality} = \frac{\text{Percent survial in the untreated controls} - \text{Percent survival in the treated insects}}{\text{Percent survival in the treated insects}} \times 100$$

The control mortality should be less than 20 per cent; otherwise, the corrected mortality will not be reliable. The mortality in control so obtained will affect the precision of the result.

5. Preparation of Stock Solutions

Technical grade (95–99% purity) insecticides are used for the laboratory tests. The active ingredient (a.i) of the insecticides varies so a 100 per cent stock solution is prepared using the correction factor (CF) below.

CF= 100% divided by % a.i of the insecticides.

i. For technical insecticide with 99.5 % a.i., CF= 100/99.5= 1.005
ii. Given the CF, compute the weight of the technical insecticide needed for preparing desired volume and concentration using the formula:

    Concentration of insecticides x volume x CF
iii. Example: To prepare 2.5 ml of 10000 μg/ml stock solution, the weight of insecticide needed will be 10000 μg/ml x 2.5 ml × 1.005 = 25,125 μg or 25.125 mg or 0.025 g.
iv. Weigh 0.025g of technical grade insecticide in a 6 ml screw cap vial using the analytical weighing balance and makeup the volume up to 2.5 ml with analytical grade acetone.

v. After preparation secure the cap of the vials with Para film to minimize evaporation. Store the prepared insecticide dilutions in a refrigerator (4°C) or freezer (preferably -20°C). Replace and dispose properly the tips after preparation of an insecticide.

6. Bioassay Methods

Bioassay methods commonly employed for insecticide toxicity evaluation are topical application, potter's tower method, injection method, dipping method (leaf dip and larvae dip), contact or residual method, film method etc. The important ones are discussed here.

6.1 Topical Application

A commonly employed method is topical application, where the insecticide is dissolved in a relatively nontoxic and volatile solvent, such as acetone and small, measured droplets are applied at a chosen location on the body surface on the thorax of individual third stage larvae with an operated micro-applicator (Durmuşoğlu *et al.,* 2015). The advantages of this method are, the high degree of precision and reproducibility that can be attained, large number of tests can be performed in a relatively short time, small number of insects (10–20) required per replication, simple and inexpensive equipment needed, small amount of toxicant and solvents used.

6.2 Potter's Tower Method

Uniform spraying or dusting on the body of insect can be done by means of Potter's Tower. Potter (Potter, 1941; Potter, 1952) designed a spray tower with a twin-fluid nozzle mounted centrally at the top of an open ended metal tube where the sprays falls vertically and deposits on horizontal plane. The topical application on entire insect body can be done by potter's tower by keeping the Petri dish containing known number of insects under the bottom part of the tower and spraying inside through nozzle fitted in the lower side by maintaining a particular pressure. This methodology simulates field exposure conditions and hence is informative for pest management. The major disadvantages of the method are the potter's tower method has a slow execution time and a high initial cost, due to the necessity of purchasing equipment.

6.3 Injection Method

In this technique the insecticide is directly injected into the insect body (thorax in insects) by hypodermic needle. The method employs a very fine stainless steel needle (27 or 30 gauge thickness and 0.41 or 0.30 mm in diameter) is used and the quantity of toxicant required is measured by micrometre. Small glass needles of 0.1–0.16 mm in diameter are used for injection in small insects. The insecticide is commonly dissolved in propylene glycol or peanut oil and injected intraperitoneally (into the body cavity). Care should be taken to avoid bleeding by the insect. This method is specifically employed to know the exact amount of toxicant needed inside the body of the insect (Sun, 1963).

6.4 Dipping Method

This method is employed when topical application or injection methods are impractical, for example: with small plant-feeding insects, stored-product insects, housefly larvae, insect eggs, red spiders, etc. The insects are picked up with a pair of forceps and dipped into an aqueous solution of the chemical for short periods of time, which is either a suspension or an emulsion (Simon, 2008). In another dipping method the leaves are dipped in aqueous solution of insecticides for definite period in serial dilution of different concentrations. The leaves were carefully drained of excess solutions and air dried for an hr before being used (Immaraju *et al.,* 1990). After few minutes the known numbers of insects watch feed on the leaves in question are released and the mortality counts are done after a definite period. This technique allows the product to be evenly distributed on the leaf surface and makes it possible to check whether or not the field doses are effective for the pest control (Siqueira *et al.,* 2000).

6.5 Film Method

The technique involves insecticide solution which is usually deposited on glass surface such as petri dish, flask, vial, wide mouth jar etc. Petri dishes are most commonly used for evaluating insecticide efficacy. In this approach petri dishes (5 cm diameter) are coated with one millilitre solution on their inner sides and the solution is allowed for uniform spreading in the petri dish by swirling it gently and then allowing it to dry up at room temperature. The target test insects are then released onto the film of the toxicant in the container. Thereafter; the known numbers of insects are exposed on for a period of 18–24 hours depending upon the recalculation (Sun, 1963; Birah *et al.*, 2008).

The susceptibility of any population to a poison is assessed by constructing a dosage-mortality curve in which the dosage is plotted against the percentage mortality at a given period of time. Such a plot produces a sigmoid curve whose asymptotic approaches (infinite ends) at the regions of zero and 100 per cent mortality are difficult to define without extensive testing. In probit transformation, the sigmoid curve is converted to a straight line by plotting the logarithm of the dosage against the probit value of percent mortality. This method of computation yields a straight line, which greatly facilitates the determination of the LC50, LC90 or any other lethal concentration/dose.

7. Utility of Bioassay

1. Bioassay is adopted to ascertain the potency of the chemicals as insecticides.
2. It helps to find out the property of synergism, potentiation and antagonism of a compound when used in a mixture with an insecticide.

3. Comparative or relative toxicity of insecticides is worked out on the basis of LC50 obtained as a result of bioassay. This gives an index for selecting promising insecticides for field trial against insect pests.
4. Bioassay helps in evaluation of insecticides for their safety to pollinators, predators and pathogens.
5. The quality of marketed insecticides can be checked through bioassay of samples collected and comparing them with standard.
6. Through bioassay lethal time LT50 required to kill 50 per cent population to test animal. ED50 or EC50 i.e. the dose or concentration of chemical brings out sterility or other quantitative effects in 50 per cent population can also be worked out.
7. Bioassay can also be utilized for estimation of micro quantities i.e. residues of insecticides in different commodities in order to alarm the consumers from hazards associated.

8. Limitations

1. Requirement of most sensitive organism for particular toxicant.
2. Rearing, handling and maintaining of uniformity of test organism. There may be the complexity of rearing or assay method for a specific organism.
3. Standardization of observation time.
4. Does not tell about the quantities of the different toxic metabolites in case of residue determination.

9. Conclusion

The understanding of median lethal dose, lethal concentration and toxicity is very important for the better evaluation of the toxic of the particular pesticides. Bioassay can be a useful tool for the determination and study of different agricultural pesticides. It can be simple, swift, versatile and highly sensitive to a wide range of toxicants. Generally little or no expensive equipment or highly trained help is required. Although used to best advantage when the toxicant is known, bioassay can be used in some instances to identify toxicants.

10. References

Abbott, W.S. (1925). A method of computing the effectiveness of an insecticide. *J. econ. Entomol*, *18*(2): 265–267.

Bacci, L., Picanco, M.C., Barros, E.C., Rosado, J.F., Silva, G.A., Silva, V.F. and Silva, N.R. (2009). Physiological selectivity of insecticides to wasps (Hymenoptera: Vespidae) preying on the diamondback moth. *Sociobiology*, *53*(1): 151–167.

Ball, H.J. (1981). Insecticide resistance — a practical assessment. *Bulletin of the ESA*, *27*(4): 261–262.

Birah, A., Mahapatro, G. and Gupta, G. (2008). Toxicity evaluation of emamectin benzoate against tobacco Caterpillar (spodoptera utura) by three different assay techniques. *Indian Journal of Entomology, 70*(3) : 200–205.

Burchfield, H.P., Hilchey, J.D. and Storrs, E.E. (1952). An objective method for insecticide bioassay based on photomigration of mosquito larvae. *Contrib. Boyce Thomson Inst, 17 :* 60.

Chhillar, B.S., Gulati R. and Bhatnagar, P. (2007). Bioassay In: Agricultural Acarology. Daya Books, pp. 220–230.

Dewey, J.E. (1958). Utility of bioassay in the determination of pesticide residues. *Journal of Agricultural and Food Chemistry, 6*(4): 274–281.

Durmuşoğlu, E., Hatipoğlu, A., Gürkan, M.O. and Moores, G. (2015). Comparison of different bioassay methods for determining insecticide resistance in European Grapevine Moth, Lobesia botrana (Denis & Schiffermüller) (Lepidoptera: Tortricidae). *Turkish Journal of Entomology, 39*(3): 271–276.

Finney, D.J. (1971). *Probit Analysis: 3rd Edn.* Cambridge University Press.

Immaraju, J.A., Morse, J.G. and Brawner, O.L. (1990). Evaluation of three bioassay techniques for citrus thrips resistance and correlation of the leaf dip method to field mortality. *J. Agric. Entomol, 7*(1): 17–27.

Paramasivam, M. and Selvi, C. (2017). Laboratory bioassay methods to assess the insecticide toxicity against insect pests-A review. *J. Entomol. Zool. Stud, 5:* 1441–1445.

Potter, C. (1941). A laboratory spraying apparatus and technique for investigating the action of contact insecticides. With some notes on suitable test insects. *Annals of Applied Biology, 28*(2): 142–169.

Potter, C. (1952). An improved laboratory apparatus for applying direct sprays and surface films, with data on the electrostatic charge on atomized spray fluids. *Annals of Applied Biology, 39*(1): 1–28.

Regupathy, A. and Dhamu, K.P. (1990). *Statistics work book for insecticide toxicology.* Suriya Desktop Publishers.

Simon J.Y. (2008). The Toxicology and Biochemistry of Insecticides. CRC press, pp. 1–276.

Siqueira, H.Á.A., Guedes, R.N.C. and Picanço, M.C. (2000). Insecticide resistance in populations of Tuta absoluta (Lepidoptera : Gelechiidae). *Agricultural and Forest Entomology, 2*(2): 147–153.

Sun Y.P. (1963). Bioassay-Insects. In: Analytical methods for pesticides, plant growth regulators and food additives vol.–1. (*Eds. Gunter Zweig*), Academic Press; New York; San Francisco; London, pp. 399–423.

Practical – 10

1. Objective

Bioassay of Fungal Entomopathogens against Grass hoppers or Locusts.

2. Why Do Bioassays?

a. To find new isolates of fungal pathogens which could be used in biological control of insect pests.

b. To compare new isolates with the ones we already know about.

c. To assess the virulence of new isolates against other insect pests.

d. To check the virulence of spores from mass production.

3. Insect Bioassay

Laboratory bioassays of insects now days became increasingly crucial because of their predictive value and generating comparative toxicity data on many chemicals in relatively short times, at relatively low expense. The laboratory investigations provide a better understanding of insect-insecticide or insect-plant-insecticide interactions. It is a simple, versatile, easy and sensitive technique for determining toxicity of wide range of chemicals, which greatly facilitates the determination of the LD50, LC50 or any other lethal concentration/dose (Burchfield, 1952).

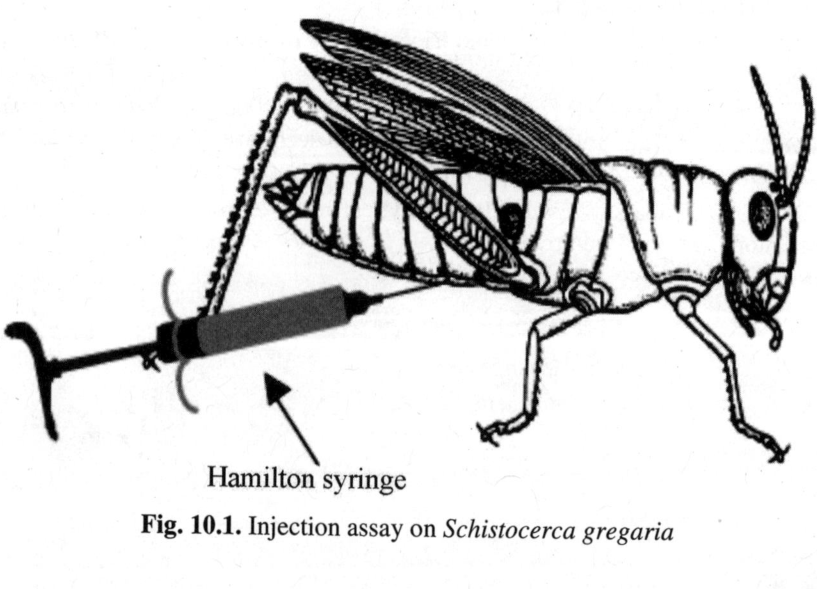

Fig. 10.1. Injection assay on *Schistocerca gregaria*

Before you begin make sure that the insect colony is free of protozoan infections, such as *Malamoeba* and *Nosema*. Some insects, in particular *Schistocerca gregaria* survive well in individual plastic boxes with no food. Other species may need to be kept in cages and provided with food.

4. Material required

a. You will need equal numbers of adults of both sexes, 9–12 days post fledgling (not sexually mature).

b. Take plastic food boxes (not airtight).

c. Put a piece of dry tissue paper in the bottom of each box - nothing else.

d. Put one insect in each box.

e. Finish handling the insects before you touch the fungus.

f. If you are using an un-inoculated control, set up the control before you touch the fungus.

g. After handling fungus DO NOT enter the insect room again on the same day.

4.1 Spore Preparation (*Metarhizium* and *Beauveria* sp.)

a. Use SDA slopes in 25 ml Universal bottles.

b. To inoculate, streak the agar liberally and evenly with conidia of the fungus to be tested.

c. Incubate at 22–25°C in 12 hours artificial light and 12 hours darkness.

d. Harvest conidia 10–30 days after inoculation.

(Note: It is a good idea to choose a standard age e.g. 15 days)

e. To harvest conidia: add 10 ml good quality oil (groundnut or soya oil, without added antioxidants or fungicides) to each bottle.

f. Use a paint brush or spatula to brush the conidia into the oil.

g. Sieve the suspension through a 75 micron sieve (this removes hyphal fragments).

h. Agitate the suspension in a bath sonicator for three minutes or a probe sonicator for 1 min. If the conidia are still clumped agitate for a further two minutes.

(Note: If you do not have a sonicator, add glass beads to the spore suspension and use a Vortex/Whirl mixer or shake the bottle vigorously).

i. The water in the suspension will settle to the bottom. Remove as much of the water as you can by decanting.

j. Use the formulation on the same day.

k. If you cannot use the formulation on the same day, you must add dry, non-indicating silica gel to absorb water.

   (Note: You must remove all surplus water to prevent conidia germinating)

l. Check conidial concentration using a haemocytometer.

m. Calculate the dilution needed to get 5×10^7 spores/ml.

n. Use two microliters 5×10^7 spores/ml suspension, this is equivalent to 10^5 spores for each adult locust (approximate weight 2 g). Reduce the number of spores for smaller insects.

4.2. Bioassay methods

Bioassay methods commonly employed for insecticide toxicity evaluation are topical application, potter's tower method, injection method, dipping method (leaf dip and larvae dip), contact or residual method, film method etc. The important ones are discussed here.

5. Inoculation

a. For each strain use 25 insects.

   If you are screening new isolates you do not need to replicate the assays. To replicate assays, use 3×20 insects.

b. Large grasshoppers: to inoculate use a micropipette to place 2 µl of suspension behind the pronotum.

   You can hold the pronotal shield forward with a blunt needle or small spatula, but be careful not to damage the cuticle.

   (N.B. Inoculating with 2µl of 5×107 conidia/ml gives a dose of 100,000 conidia per insect (105 spores/insect).

c. Put the insect back in the box.

d. Store at 30°C in 12 hours light and 12 hours dark.

e. Finally, check the viability of spores in each spore suspension used by spreading 0.1 ml of the spore suspension onto an SDA plate as described in the section on Germination testing in this section.

6. Recording

a. Assess mortality every day.

b. If you find dead insects, look for a red colouration, this often is indicative of death due to infection by *Metarhizium* and *Beauveria* sp. Death caused by these fungi also results in the insect cadaver becoming rigid. Insects killed by other causes such as bacterial infection normally appear limp and the limbs will appear loose. A black colouration under the thorax or abdomen of the dead insect is also indicative of bacterial infection.

c. Check your diagnosis, by saturating the tissue paper in the box with water and incubate the dead insect for a few days in the box to see the growth of fungus on the outside of the insect.

d. Even if a fungus does not develop on incubation, it may still have been the cause of death *e.g.* by toxins produced by the fungus. Compare with a blank formulation control to see whether the pathogen did cause death or whether there was some other cause. If the controls are healthy, then record all mortality as caused by the fungus. Record assays for 12 days. For the final result calculate each day as cumulative percentage mortality over the assay period (Chaieb *et al.*, 2007).

7. Analysis

a. Plotting cumulative mortality curves give a basic comparison between strains and you can read LT50s from these lines.

b. Calculate the average and the variance in survival times so that you can make quantitative comparisons between strains.

c. For a more detailed comparison, use a range of doses and calculate the LD50 by probit analysis.

d. Specialised software such as the MELTIMOR programme, or the Kaplan-Meier survival analysis in SPSS should be used to analyse the data.

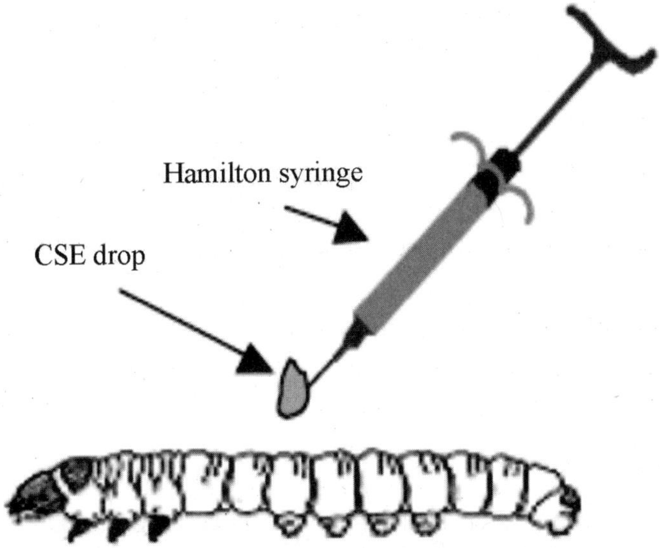

Fig. 10.2. Topic application assay on Lepidopteran larvae

8. Bioassays against Other Insects

a. It is necessary to develop a bioassay protocol for any new insect species; normally an oil drop bioassay gives the most reliable results, but it is necessary to find the best place for the oil droplet.

b. It is also important to ensure that the insects are kept under ideal conditions, with very low control mortality.

c. Always select a standard strain which can be included in all bioassays for comparison.

d. Check each new species to verify that it is not affected by a control inoculation with blank oil (Chaieb *et al.*, 2007).

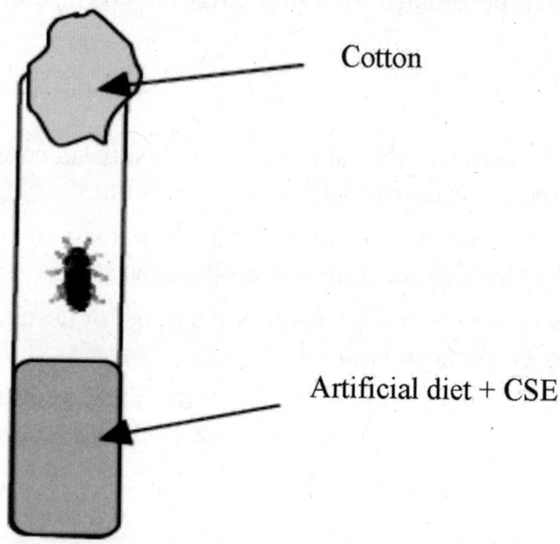

Fig. 10.3. Diet treatment assay on weevils

9. Limitations

a. Requirement of most sensitive organism for particular toxicant.

b. Rearing, handling and maintaining of uniformity of test organism. There may be the complexity of rearing or assay method for a specific organism.

c. Standardization of observation time.

d. Does not tell about the quantities of the different toxic metabolites in case of residue determination.

10. References

Burchfield, H.P., Hilchey, J.D. and Storrs, E.E. (1952). An objective method for insecticide bio-assay based on photomigration of mosquito larvae. *Contributions from Boyce Thompson Institute.* 17: 57–86.

Chaieb Ikbal, Ben Halima Mounia and Ben Hamouda Habib (2007). Toxicity Experiments of the Saponic Extract of *Cestrum parqui* (*Solanaceae*) on Some Insect Spices. *Journal of Entomology,* 4: 113–120.

Paramasivam, M. and Selvi, C. (2017). Laboratory bioassay methods to assess the insecticide toxicity against insect pests - A review. *J. Entomol. Zool. Stud*, 5 pp. 1441–1445.

■■■■